SCIENCE 101

ECOLOGY

HarperCollins books may be purchased for educational, business, or sales promotional use. For information, please write: Special Markets Department, HarperCollins Publishers, 10 East 53rd Street, New York, NY 10022.

Produced for HarperCollins by:

Hydra Publishing
129 Main Street
Irvington, NY 10533
www.hylaspublishing.com

FIRST EDITION

The name of the "Smithsonian," "Smithsonian Institution," and the sunburst logo are registered trademarks of the Smithsonian Institution.

Library of Congress Cataloging-in-Publication Data

Freeman, Jennifer, 1961-
  Science 101: Ecology / Jennifer Freeman. -- 1st ed.
      p. cm.
  Includes bibliographical references and index.
  ISBN: 978-0-06-089133-6
  ISBN-10: 0-06-089133-5
  1. Ecology. I. Title.

  QH541.F697 2007
  577--dc22

                          2007045478

  07 08 09 10   QW   10 9 8 7 6 5 4 3 2 1

SCIENCE 101

ECOLOGY

Jennifer Freeman

Collins

An Imprint of HarperCollins*Publishers*

CONTENTS

WELCOME TO ECOLOGY

Left: A tree and other plant life thrive even in an arid desert region. Despite severe conditions, plants and wildlife find ways to draw the energy they need to survive. Top: A lush mountainous area of Hawaii, the product of much sunlight and water. Ecosystems like this one, and all around the world, continue to mutate and evolve. Bottom: A lone bird flies over the frigid Arctic terrain. Scientists monitor the Arctic closely to gauge global warming and its effects on climate.

The closer we look at the web of life on Earth, the more amazing it seems. Dragonflies, antelope, and oak trees share a basic cell structure. It is a wonder that, made up of the same building blocks, they evolved such wildly varied shapes and live such different lives.

On the Earth's surface, in coastal, forest, and desert communities alike, the Sun's energy provides the fuel. In vibrant communities around deep sea vents, chemicals are the basic fuel of life. Even deeper, in rocks beneath the ocean, bacteria use radioactivity from Earth itself to stay alive. Bacteria as a group are extremely diverse. Some are able to live on acids, others need no oxygen.

Creatures and environments are fantastically diverse, yet each one strives to find the energy it needs to survive and to pass its genes to the next generation. Not all succeed. If they did, the world would be rapidly overpopulated by all manner of living things. From simple beginnings, life has evolved into a tremendous array of creatures. There are more than 17,000 species of butterfly alone. A student of ecology cannot help but gain respect for the planet Earth, where simple molecules have been spun into such a remarkable web of life.

SCIENCE OF ECOLOGY

Through his research, ecologist James Lovelock found that the atmospheres of Mars and Venus, dominated by carbon dioxide, were nearly inert, close to a state of chemical equilibrium. Earth's atmosphere, in contrast, was full of gases that can react with one another, a state of deep chemical imbalance, or disequilibrium.

The Earth's systems are regulated through physical, chemical, geological, and biological processes. Understanding the mystery of Earth's intricate processes, what they are and how they work, is what the science of ecology is all about.

This book, *Science 101: Ecology*, introduces the basic ideas of ecology. We will look at why there are fewer coyotes on Earth than ants. We will consider how lizards have evolved behaviors to balance the energy spent on basking and hunting; look at the microscopic communities in healthy soil; learn how algae relate to dairy farms, what influences peacocks' choice of mates, what portion of Earth's species are beetles, and how each of these fits in with the theories of ecology.

We will explore how animals' behavior, such as the way wolf parents raise their young and the construction of birds' nests, evolves from interactions with the environment. We will learn about life within the heated rocks of the Earth's crust.

CHANGING TIMES

We live in a world where for billions of years organisms have survived and evolved, affected or been affected by their environments. They have thrived when competition was scarce and food plentiful, died off when carrying capacity was exceeded, and relied on genetic diversity to get through tough times.

We live in an age when the human species has become increasingly dominant in most of the planet's habitats, influencing basic elements

Left: Cows bred for both milk and food stand in the shadow of a power station in England. Industrial progress has often negatively affected global ecology. Right: A section of tropical rain forest. Rain forests help rid the Earth of carbon dioxide and maintain a balanced climate.

A range technician takes samples of larkspur in Utah. Plant samples help agricultural researchers understand such things as soil content and moisture levels.

of ecosystem function. The exploding human population, and the resources humans consume, threatens as many as a quarter of all species currently living on Earth with extinction.

> "Unraveling the mysteries of Earth's intricate processes, what they are and how they work, is what the science of ecology is all about."

Human activity reaches all corners of the world, even affecting the atmosphere itself. Yet the last remaining wild lands on Earth still provide inspiration and meaning, a sense of hope and opportunity. The combination of ecology with special concern for the expanding influence of humans has led to the field of environmental studies.

A butterfly fish swims along a reef as the Sun shines through the water's surface. For ocean environments, chemicals, rather than sunlight, provide energy to sustain life.

While ecology focuses on a basic understanding of how individuals, populations, communities, and ecosystems interact, environmental scientists seek to apply that knowledge to solve problems created by humans.

If this book gives readers a greater understanding of the planet where we live, and how life depends on the processes that keep the Earth in balance, perhaps it will inspire them to make choices that will help keep the dominant species, humans, on a sustainable path.

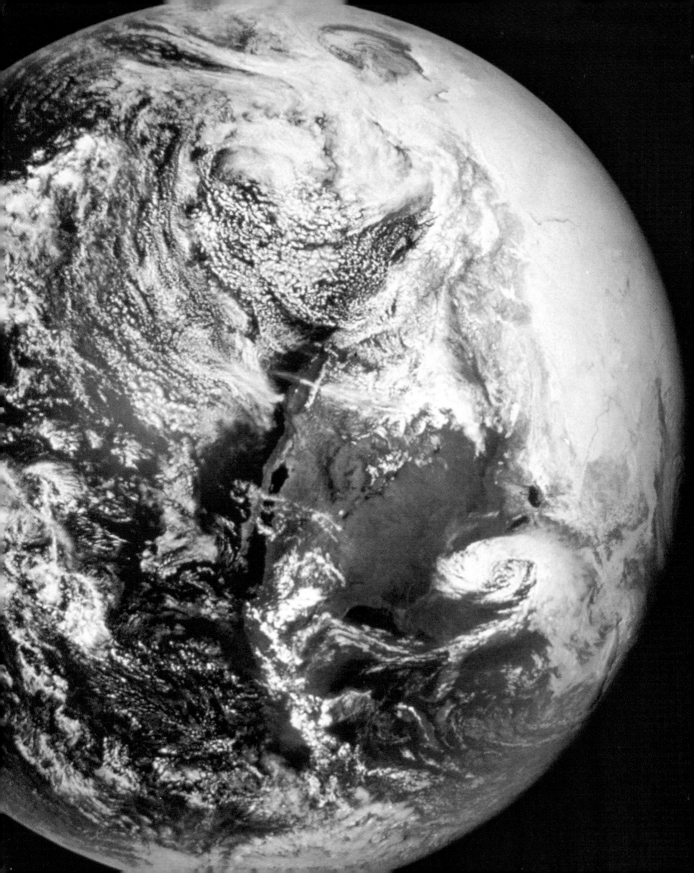

CHAPTER 1

LIFE ON A SMALL PLANET

Left: The planet Earth, viewed from space, inspired astronaut Loren Acton to write: "There, contained in the thin, moving, incredibly fragile shell of the biosphere is everything that is dear to you." Top: A tropical rain forest in Brazil. Bottom: A canyon in Egypt's Sinai Desert. The lush rain forest environment contains many times more animal and plant species than an arid desert ecosystem.

Ecologist Edward O. Wilson has said humanity is now "passing through a bottleneck," a period of great challenge. There are now so many humans, using so many of the Earth's resources, that humans have the power to alter the very life support systems of our planet. How will a growing population that will reach 9 billion souls live and eat without destroying the living planet?

The science of ecology essentially is the basic quest to determine how the natural world works. Ecology is a relatively new science. It is about understanding how the parts fit together: predators and parasites, plants and soil, geography and genetics, insects and bacteria. It addresses key questions: Why are there so many species in the tropics? How do desert animals manage water resources? It also addresses applied questions: What will happen to life on Earth and the cycles that support it as climate-warming gases build up in our atmosphere? How is life on Earth affected by habitat loss or damage to the ozone layer? These are just a few of the issues studied by ecologists. The skills needed to survive this time of challenge must include a sound understanding of what makes the planet live in the first place.

Land of Opportunity

In the nineteenth century, the wild lands and wildlife of the United States seemed boundless to recently arrived Europeans. Railroads bored through the center of the country. Settlers plowed the prairie under for corn, and fenced off wild land for farms and ranches. The race to draw riches from the new land—furs, gold, timber, beef—rose to a fever pitch.

Buffalo were slaughtered so fast that where tens of millions had blackened the prairies in 1850, by 1884 their numbers had been reduced to fewer than one thousand. Egrets were hunted nearly to extinction as well, their elegant feathers prized for ladies' hats. Beaver, too, were hunted to a tiny fraction of their former population, their fur made into top hats. Ancient redwoods were chopped down for lumber to build gold-rush boomtowns.

Forest was cleared, towns and cities constructed. People, plants, and animals that had lived on the land for thousands of years were replaced by the newcomers.

APPRECIATING NATURE

In the middle of this race for riches, another kind of voice emerged. At the same time that buffalo and egrets were brought to the brink of extinction, some Americans began to write and

speak of the majesty of nature. The land and its wild creatures, they argued, should be cherished as more than just a source of raw materials for building houses and fortunes.

"Everybody needs beauty as well as bread, places to play in and pray in, where nature may heal and give strength to body and soul," wrote Scottish-born naturalist John Muir. Some saw that nature was under threat from the great changes happening as

Above: An elk in the North American Rockies. To nineteenth-century Americans the idea of preserving rather than exploiting untapped land and wildlife was revolutionary. Top left: Shooting buffalo from the Kansas-Pacific Railroad line.

the country grew and expanded westward. Preserving and protecting nature was necessary, the new argument went, not only for conserving resources, but for the very health of the human spirit.

Of course, the view that Earth is a living thing to be nurtured and honored by man was common among Native Americans. Iroquois law in the nineteenth century, for instance, stated: "In our every deliberation, we must consider the impact of our decisions on the next seven generations." But for European Americans, the idea seemed new.

A NEW SCIENCE

While naturalist writers wrote of the beauty and wonders of nature, others began to create a new, scientific understanding of the landscape. Earlier students of Earth and its life had focused on the separate pieces they saw—flowers, microbes, field mice, weather. But by the late nineteenth century, scientists were looking to understand how natural systems function over time and space, what makes species succeed or fail, and how living and nonliving worlds connect.

Scientists began looking at nature as a series of interconnected systems, small units and large ones, all part of the bigger picture. They began to research

A waterfall in Yellowstone National Park, Wyoming, one of the earliest preserved wild lands in the United States, established as a public park in 1872.

NATURALISTS PROMOTE WILDERNESS PARKS

One of the first European American voices for wilderness preservation was transcendentalist author Henry David Thoreau, whose 1854 book *Walden* poetically conveyed the great spiritual value of nature. Thoreau wanted all Americans to share his passion. He wrote, "Each town should have a park, or rather a primitive forest . . . a common possession forever, for instruction and recreation."

Scottish-born naturalist John Muir was even more ambitious than Thoreau. Muir thought large areas of wilderness should be set aside and preserved as national parks. Muir especially loved California's Sierra Nevada, where stands of towering redwood trees thousands of years old were being rapidly chopped down for timber. Muir helped persuade President Theodore Roosevelt to preserve America's wild heritage in the first national parks.

Above: John Muir and John Burroughs were naturalist writers and friends who inspired Americans to appreciate wild lands. Below right: Stereoscope image. John Muir, standing at right on Glacier Point, Yosemite Valley, California, inspired President Theodore Roosevelt (left) to found the United States' first national parks.

the dynamics of nature's chemical, physical, and biological cycles, and to recognize that the success of all creatures depends on their interactions with one another, with other species, and with the nonliving components of their home environments.

This new science was dubbed "ecology," *ology* meaning "study," and *eco* from the Greek word *oikos*, meaning "the place where we live, our home."

Rare Mammals

It is easy for humans to feel that we are the central and most important species on Earth. After all, it is human faces we see in the mirror each day, humans we love, human communities we inhabit.

Looking at species on Earth from a numerical point of view, however, a completely different picture emerges. Defined by the number of species on Earth, humans, along with all other mammals, are rare indeed.

Scientists have described approximately 1.4 million species of living organisms. This number includes all manner of life, ranging from bacteria to oak trees to lions. Of these, almost two-thirds are insects; another quarter million or so are plants. A mere 4,000 known species are mammals.

PLENTIFUL INSECTS

More than 750,000 of the 1.4 million known species on our planet are insects. The insects, which include beetles, butterflies, ants, and termites, far outnumber their vertebrate cousins.

The most common type of insect is the family known as coleoptera, or "sheathed wing" insects—the beetles. There are nearly 300,000 known species of beetle, more than all noninsect animal species combined.

A story is told about J. B. S. Haldane, a well-known British

Above: A ladybug diligently explores a wheatstalk. Top left: A lioness hungrily prowls the African savannah. Mammals such as the lion are relatively rare in comparison to insects such as the ladybug. More than half of all species on Earth are insects.

biologist and evolutionary thinker. Once, Haldane was asked what a person could conclude as to the nature of the Creator from a study of his creation. Haldane is said to have answered, "an inordinate fondness for beetles."

Insects play many important roles. Bees and butterflies are famous for helping plant life to reproduce by pollinating flowers.

Flies and their larvae help to break down dead plant and animal matter so that it can be recycled. Grubs underground improve growing conditions by bringing air into the soil. Ladybugs support plant life by controlling the populations of plant-sucking aphids. If there were suddenly no insects on Earth, most plant and animal life would be extinct within a year.

Decomposers are organisms that digest or break down living material that has died. Decomposers, such as this mold, fill crucial ecological roles, breaking down dead organic matter and returning the nutrients to a form that can be recycled by new life.

MICROBES THRIVE

In a list of the species that are most important to life on Earth, near the top should be the fungi and bacteria, the ever-present microbes that live on or in almost every material and environment on the planet.

Animals and plants could not survive long without fungi and bacteria. In getting nourishment for themselves, these organisms turn the matter that life is made of back into its basic particles and compounds, so that they can be recycled and used again by new life. These decomposers are helped along by many types of beetles, ants, and termites. Some live on fallen wood, others decompose dead animal bodies, fallen fruit, or feces. Their main food is material that other living things no longer need, also known as detritus. For this reason the recyclers are known as detritivores. Rotting animal carcasses swiftly absorbed into the soil, fallen fruit moldering in an orchard, and dead tree trunks dissolving on the forest floor all are signs that the decomposers and detritivores of the world are performing their necessary work.

Like mold and bacteria, certain species of beetle, such as this dung beetle, are decomposers. They help turn dead plants and animals back into their chemical components.

The Plant Kingdom

Plants are multitalented organisms. They exhale the oxygen that animals need for life. They store carbon and water, helping to regulate the Earth's climate. But perhaps plants' greatest ability is capturing the Sun's energy and producing food from it. In the ocean, phytoplankton and macroalgae perform this job.

Almost all energy used by life on Earth originates from the Sun. Plants use photosynthesis (*photo*

means "light," and *synthesis* means "combining") to produce food energy from light.

Plants store energy in their leaves and stems, and roots and seeds. Because plants produce energy from sunlight, they are called primary producers. Autotrophs, which means "self-feeders," is another term for organisms that can feed themselves using only nonliving (abiotic) ingredients such as water, sunlight, and nutrients.

In the rain forest, many creatures never come down to ground level. Epiphytes, types of plants that include orchids, get all the materials they need from the rainwater, air, and sunlight they find on the branches of

big trees. Their roots never even touch soil. In the oceans, macroalgae and tiny organisms called phytoplankton take on the job of photosynthesis, along with a few marine plants.

ENERGY FROM THE SUN
Plants, the producers of energy, directly or indirectly feed all other creatures. Herbivores, who live by consuming plants, are called primary consumers. Carnivores, feeding on creatures who have eaten plants, are called secondary consumers. Mammals, such as humans and raccoons, who can eat both plants and meat are called everything eaters, or omnivores. Any creature that eats plants, whether

Above: An epiphyte-covered tree trunk in a Colombian rain forest. Right: Epiphytes such as this orchid occupy a niche high in rain forest trees, absorbing nutrients from rain, air, and the Sun. Top left: North American temperate forest.

A herbivore is an animal that gets its energy from a diet of plant food only. Herbivores such as this orangutan, shown eating grass, are known as secondary consumers. All organisms, from top predators to moss, ultimately get their energy from the Sun.

grass, nuts, or fruit, or that eats another animal that eats plants, is ultimately getting its energy from sunlight. An orangutan that eats figs and leaves is eating the stored energy of fig trees. The orangutan is a primary consumer.

Secondary consumers include birds that eat grubs and insects, and spiders that feed on flies that become tangled in their webs.

An ocelot that eats a coatimundi that ate beetles that ate berries is called a tertiary consumer. But ultimately the ocelot is getting its energy from the sunlight, too.

PRODUCTIVITY CHAMPIONS

The dry weight of plant and animal life that is created in a place is called biomass. The rate at which biomass can be created

with the light, water, and nutrients of one location is called the productivity of that place.

Tropical rain forests are champions of productivity. Trees, birds, animals, fungi, molds, bacteria, insects—a diverse sampling of life—utilize most of the nutrients and moisture in the system, leaving little in the soil. Rain forest soil is very poor. Nearly all nutrients needed by the living organisms are continually recycled through the plants and other living organisms. Even water is recycled: studies have shown that rain forests near the equator make their own rain by transpiring (evaporating through their leaves) three-quarters of the rain that falls on them, so that it can go up and fall again.

Biomass can be measured in a number of ways. For example, an ecologist might weigh all living material in a given area or count and size individual organisms. Ecological scientists have also developed computer models to make estimates of biomass. In order to test the accuracy of the models, scientists compare results from the hand-collected samples to the computer-generated estimates.

Bacteria probably account for a greater amount of biomass than all other organisms combined. Plants and algae, both primary producers, come in second, followed by insects and fungi. At the apex of the pyramid are all the birds and mammals, from hyacinth macaws to golden lion tamarins.

Energy and Order

Primary producers, including plants, algae, and cyanobacteria (a major group of photosynthesizing bacteria), capture only 10 percent of the energy that radiates to Earth from the Sun. Most of the Sun's energy goes into the air or ocean, or reflects back into space. Plants store the energy that they capture through photosynthesis as sugar or glucose molecules, waiting to be turned into energy when they are "burned" for fuel.

When a caterpillar eats a leaf, it captures only a fraction of the Sun's energy from the plant. In turn, the flycatcher that eats the caterpillar gets only a small amount of that energy from the caterpillar. The remaining energy will have been lost to heat, reproduction, and waste material generated by the caterpillar.

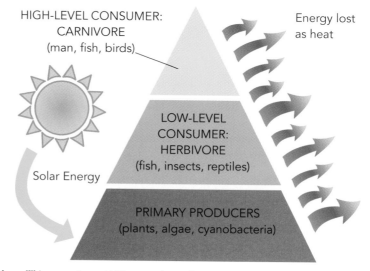

Above: This energy pyramid illustrates how solar energy is absorbed by primary producers and transferred to higher level species. Top left: Mangrove tree in the Florida Everglades.

This is the energy pyramid, a concept that illustrates how energy is transferred from one level to another as animals of the upper level prey on those of the lower level. At each level, organisms lose a significant amount of energy as heat, and pass less of the Sun's energy up to the next level. Because animals cannot use sunlight to make their own food, they need to feed on plants and on other animals for energy.

LEVELS OF ORDER

The field of ecology approaches nature's systems by dividing them into different levels. The first level is the individual organism, its physiology and behavior, what it eats and where it lives. For instance, the survival of the mangrove tree that lives in coastal wetlands is dependent on many factors, including its

A monarch caterpillar eats a milkweed leaf. The milkweed gets its energy directly from the Sun and soil, but the herbivore caterpillar must eat large quantities of leaves to get enough energy to grow into a butterfly. Herbivores, as secondary consumers, need a lot of energy to stay alive.

River ecosystems support a variety of life both in and around the water. Rivers cycle water and nutrients through the land. Natural riverbanks help control erosion and flooding as well as provide habitat for migrating birds, which feast on the life at the water's edge.

ability to tolerate the salinity of its environment. The second level is populations of organisms—individuals of the same species living together and interacting, such as mangrove trees that together may trap sediments to create a stable environment, or that individually may compete for space and light.

Level three is communities, which include populations of various species living together or interacting with one another, from mangroves to shrimp to bass, alligators, manatees, and egrets, all in one location.

Level four is ecosystems, which include the nonliving (abiotic) elements of the environment, as well as its living (biotic) organisms. The mangrove swamp ecosystem is comprised of a large number of interdependent elements, including fresh and salt water, climate, sediment, and pollution from agricultural runoff in addition to organisms such as the shrimp and alligators mentioned above.

Level five is the ecoregion, which considers nature on the scale of large landscapes, such as tropical rain forests.

WETLANDS ECOLOGY

Wetlands have a very high rate of productivity, nearly as high as rain forests and much greater than agricultural lands. Wetlands serve important ecological functions as nurseries for marine, terrestrial, and freshwater animals, including crabs, snails, worms, fish, birds, frogs, and other amphibians. About half of the world's migratory birds nest in or near wetlands because of the rich food sources they provide.

Wetlands also protect coastlands from flooding and hurricane damage and help with pollution control by filtering large quantities of water.

Many wetlands have been drained and turned into farmland, or drained for housing and other structures. Because they are too watery for humans to walk on, wetlands are often dismissed as wasteland; the draining of wetlands is sometimes referred to as "land reclamation." About half the wetlands in North America have been lost. Not coincidentally, about half of all endangered species listed in North America depend on wetlands for survival.

Danger of Extinction

On Earth today animal and plant species are becoming extinct at about one thousand times faster than fossil records indicate is the normal rate in history. In fact, the rate of extinctions is the highest it has been since the last great extinction at the end of the age of dinosaurs.

The main reason for the high rate of extinctions today is loss of habitat. Most species of animals live in small ranges in or near tropical climates, where conditions are richest for the development of life.

Human populations have been rising rapidly since the mid-1900s, and people in search of land for farms, ranches, and houses have chopped down more than 70 percent of the tropical forests where half the world's species live. Climate change creates an even greater challenge, especially for species with diminishing populations.

The ecologist Aldo Leopold wrote, "We abuse land because we regard it as a commodity belonging to us. When we see land as a community to which we belong, we may begin to use it with love and respect."

In the past, the Earth has taken about 10 million years to regain its previous levels of biodiversity after major periods of extinction. The question that environmental scientists are now asking is how humankind and other life forms can share the Earth's sustainability over time.

EARTH'S VANISHING VERNAL POOLS

One type of habitat that is disappearing at an alarming rate is vernal pools—seasonal wetlands that are important for the survival of many species of wildlife, especially in Mediterranean-type climates. Often surrounded by spectacular flowers in spring, these pools are home to several species that are endemic, meaning that they live nowhere else.

For example, several types of tiny shrimp and more than 60 species of plants live only in California's vernal pool ecosystems, where the eggs of some shrimp dry out in the winter and hatch when the pool forms again the next year.

Other plants and animals that have adapted to the vernal pool ecosystems grow or come around each spring to take advantage of the watery landscape. Frogs lay eggs in the pools, and the tadpoles, when grown, leave the water and move on to the land phase of their lives as the vernal pool dries up each summer. Fish cannot breed in the pools, due to the lack of a permanent source of water, but migrating birds, tiger salamanders, and other animals

Above: The California fairy shrimp, one of the species upon which migrating birds rely, are endangered due to habitat destruction. Top left: Logging is seen as a cause of habitat destruction.

The last wild Tasmanian wolf.

Rock hollow vernal pool. This seasonal pool, fed by rain or snowmelt, will support new life in springtime, and then dry up by late summer, only to reappear the next year.

depend on the vegetation and shrimp for food. In recent years, less than 30 percent of California's vernal pool ecosystems remain. Many are on land that people want to use for other purposes. Conservation organizations and the government of California are working to understand and manage remaining wild landscapes to protect vernal pools and other fragile ecosystems.

MANAGING FOR LIFE

Today's ecosystem managers realize that in order to be sustainable over time, ecological management has to take into account the needs of all the inhabitants of the land, from humans and mahogany trees to chameleons and soil organisms.

For many, the primary goal is not to keep forests "pristine" without any human presence, but rather to establish a noninvasive and mutually beneficial balance between man and nature. For example, research

Above: The golden lion tamarin of Brazil is one of the world's most endangered mammals.

on mosquitoes and malaria now shifted focus away from killing mosquitoes—an action that would disrupt the balance of the ecosystems—and toward simply making them unable to transmit malaria to humans.

Management of the world's ecosystems will continue to be a great challenge. Understanding ecology can help bring humanity through the bottleneck of our enormous population growth while still preserving the integrity of Earth's ecosystems.

CHAPTER 2

THE UPS AND DOWNS OF POPULATIONS

Left: Honeybees in a hive. The population of bees in a single hive can number hundreds or even thousands. Top: Wildebeest in Masai Mara, Africa. Wildebeest populations are affected by rainfall, disease, and predators, as well as human land use for farming and livestock. Bottom: A Sumatran tiger. The densest tiger populations are no more than five individuals per 100 square kilometers.

Populations—groups of organisms of the same species that live in the same general area—may be tightly packed together like honeybees in a hive, or they may be sparsely scattered in the landscape, like tigers in a jungle.

Studying populations is a good way to understand how nature functions—better than researching either individual organisms or specific places. If one population (of grasshoppers, for instance) explodes, it may threaten the populations of its food supply (wild grasses and crops), which may cause problems for other species that were planning to eat the same food source. On the other hand, an excess of grasshoppers would provide a feast to its predators, and the populations of grasshoppers' predators (birds of prey, for instance) might increase.

Environmental stress, such as the kind caused by geographic barriers, drought, and invasion by another species, will test a population's ability to adapt. Under pressure from long-term stress such as global warming, populations that have broad tolerances and diverse eating habits are generally less vulnerable than those adapted to a narrow or specialized niche.

Balancing Act

The population of mice in a field is kept in check by hawks, foxes, owls, and snakes. The success of mouse populations depends on how well both the mice and the predators themselves are able to nest and breed. A pair of hawks with two fledglings in the nest will need extra mouse dinners. If all mouse predators suddenly disappeared, the population of mice in the field would suddenly rise. This is called an outbreak population. But it would still be limited by the amount of food available, such as seeds. The food supply in turn depends on the weather. Did enough rain fall to produce a large crop of seeds, or did drought make seeds scarce? And were other species that compete for the same food—birds, for example—also abundant?

Another limiting factor for the mouse population is habitat. The population will diminish if a construction project or a wildfire makes the field a difficult place for mice to live. For some species, destruction of habitat is such a big problem as to drive the population to extinction.

POPULATION CYCLES
Wild populations of many species rise and fall in cycles in response to limiting factors and success factors. A famous

Above left: Canadian lynx cub. Lynx populations grow and shrink in ten-year cycles, closely following the population patterns of hares. Above right: Snowshoe hare. Top left: A hawk surveys a field, keen eyes searching out a mouse.

example of such a cycle is the ten-year fluctuations of the relative populations of the North American snowshoe hare, and its main predator, the Canadian lynx. When hares are most abundant, the lynx litter size increases; drops in hare populations are followed by drops in lynx populations. The lynx's fortunes lag behind those of the hare by about a year.

Yet some studies have shown that in areas where lynx are absent, populations of hare still follow ten-year cycles of boom and bust. Why? When predators are absent, food supply becomes a stronger limiting factor for the hare. Hare populations will continue to

grow until there is not enough grass to feed them all, and then the population will crash.

POPULATION CRASH
The number of animals that a piece of land can sustain in the wild is called its carrying capacity. When natural controls are taken away, the results can be disastrous.

In 1944, 29 yearling reindeer were introduced on Alaska's St. Matthew Island, in the Bering Sea, as backup food for a small Coast Guard station. By 1957 the herd had grown exponentially to more than 1,300 reindeer, each one fatter than average reindeer in a domestic herd. Their robust health indicated that they were feasting on the nutritious tundra grasses, lichens, and willows. In the summer of 1963, a visiting biologist counted 6,000 reindeer on the island. Their average body size had decreased, however, indicating the reindeer had reached the limit of the island's resources. Because that winter had been particularly severe, the pressure of all the reindeer on the island's plant life could not be sustained.

All but 42 of those thousands of reindeer died. Only one male survived, and he was unable to reproduce. In the following years, the reindeer population on the island dwindled to zero. A graph of the reindeer population boom would start low and rise suddenly, making a J-curve, so named because it makes a line in the shape of the letter.

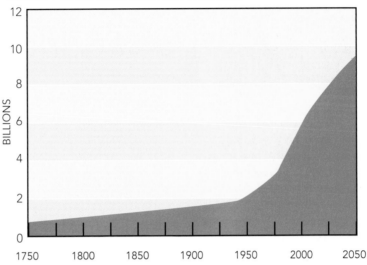

WORLD POPULATION GROWTH 1750-2050

The Earth's human population, one billion in 1800, is expected to reach nine billion in the twenty-first century. Is this exponential rate of increase sustainable?

A POPULATION EXPLOSION

The population of humans in the world was relatively stable for most of our species' history. Then, in the eighteenth century, human populations began to rise exponentially, aided by the Industrial Revolution and improvements in sanitation. The number of humans on Earth rose from about 750 million in 1750, to 1 billion in 1800, to more than 2 billion by 1950. In 2000 there were about 6 billion humans living on the planet, and the population is expected to rise to 9 billion by 2050. Increases in food production, fueled by improvements in food technology, have allowed humans to continue feeding themselves as populations steadily rise.

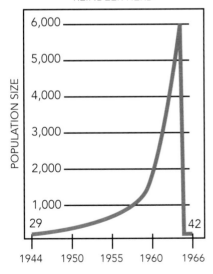

ST. MATTHEW ISLAND
REINDEER HERD

In 1963, a harsh winter killed off much of the plant life on Alaska's St. Matthew Island and caused reindeer to starve to death (right). In this graph (above), the J-curve marks the precipitous drop in the reindeer herd.

Safety in Numbers

In general, populations are more stable in the long term when the number of individuals is large. One reason for this is that a large population allows for greater genetic diversity, giving the population the ability to withstand a host of potential disasters. An outbreak of disease, an extended drought, or flash flood may kill a large percentage of a population, but if there are enough members, some—the most drought resistant, the most resourceful, or the healthiest—will survive.

And it is not just genetics that matter. In the event of a disaster that kills off a large number of individuals, a large population increases the likelihood that enough individuals will survive and reproduce successfully.

NO SAFETY IN NUMBERS
In the 1800s, billions of passenger pigeons darkened the skies of North America. In 1813, John James Audubon wrote of riding 55 miles (89 km) under a solid cloud of birds. In 1900 the last passenger pigeon was shot in the wild. From billions to extinction took about 30 years and resulted from a combination of factors that increased the vulnerability of the passenger pigeon species.

Passenger pigeons, which lived in a small number of massive flocks, were easy to hunt and reproduced slowly, bearing only one chick per year. The flocks also moved around frequently and required large, fresh tracts of forest that they virtually destroyed before they relocated.

Faced with abundant forests that were chopped into checkerboards for farmland, and rapacious hunters who harvested hundreds of thousands of birds a day, the pigeons ran out of strategies. Human competition for the birds' habitat and massive hunting played a large role in the birds' rapid extinction, but ultimately the pigeons failed because of an inability to adapt when circumstances changed.

SUCCESSFUL AS A FOX
There was a time when it seemed that the red fox would enter the list of endangered species. Like the passenger pigeon, the fox was hunted and its habitat was threatened. With houses and farmland expanding through the woodlands of North America, it seemed likely the fox would run out of woods in which to dig its den.

But the fox has proved remarkably adaptable. Foxes have made themselves at home in the suburbs. They have been seen to nest in old drainpipes and abandoned buildings if no woods are available. And if field mice or rabbits

Above: Loss of trees to farmland is one reason passenger pigeons became extinct. Top left: A stuffed passenger pigeon.

cannot be found, foxes will eat rats or even old sandwiches out of the trash. Because of this successful adaptation, the population of red foxes has remained healthy even as their former woodland habitat has diminished.

VULNERABILITY
Unfortunately, most species are not as adaptable as the fox. When the original habitat of a species is changed or made smaller, or otherwise compromised by

The Monteverde Cloud Forest reserve in Costa Rica. The multitude of species inhabiting lush, moisture-rich tropical cloud forests are vulnerable to extinction because these ecosystems are small, fragile, and threatened by climate change.

human development, the risk of extinction increases. It becomes harder for animals of the species to find suitable mates, or disperse seeds widely. Consequently, genetic pools become smaller.

Creatures that live in small, specialized niches have the hardest time adapting. For example, some species of beetles, spiders, and ants living in the rain forest canopy are adapted to a single species of tree. As rain forest is cut down, those populations have a difficult, if not impossible, time finding new places to live.

Similarly, the golden toad lived only in the Monteverde Cloud Forest in Costa Rica, a small mountaintop habitat. The cloud forest depended on cool, wet conditions, but rising global temperatures warmed the area. Then a severe drought struck in 1986–87. As the cloud forest warmed, lowland species moved up the mountain slope, and the golden toad's territory was under threat. The golden toad was last seen in 1989, and is now presumed extinct.

A European red fox raids an overturned garbage can. Foxes have proved clever at successfully adapting to life near humans.

The golden toad was last seen in Costa Rica's Monteverde Cloud Forest in 1989.

Stress on the Margins

Every population has its sweet spot. Under the right circumstances—abundant food, relative safety from predators and disease, favorable conditions for breeding or seed dispersal, and the like—the population will thrive. Thriving, it will grow, and perhaps spread to new territory.

Outside of this sweet spot, conditions are not optimum. Some species cannot exist where the water is too acidic, the soil too dry, or the air lacks humidity. These conditions may make it difficult for a population to survive.

Limiting factors—conditions that adversely affect a population's ability to sustain life—may loom larger at the margins of a population's range. The population is therefore likely to be less dense. Perhaps the climate is not optimal for the food supply, so food is difficult to find. For example, the sugar maple tree is now widespread in the eastern United States into southern Canada. In the southeast part of its range it does not reproduce as well, flowering only when it is more than 20 years old. It thrives in soils that are found in temperate climates. But climate studies predict that as the weather in the United States grows warmer in coming decades, the range of the sugar maple will shift even farther north in Canada.

SELF-REGULATION
In Europe, roe deer populations will only expand until the population reaches a certain density before numbers stabilize. Some populations of roe deer seem to have fewer offspring as resources become scarcer. North American

Above: Hunting has culled the number of wolves (right), the mule deer's (left) natural predator. The resulting change in the mule deer's ecology has spurred a boom in its North America population. Top left: The range of sugar maples shifts northward as the climate warms.

The Pacific Rim National Park in British Columbia, Canada. Although a mature forest ecosystem exists in a relatively balanced and stable "climax state," it is still in constant flux over time. Nature constantly shifts as living and nonliving forces react to one another.

mule deer, on the other hand, did not evolve such a self-regulating mechanism. Without predators to keep their populations under control, mule deer will reproduce until there are so many deer that food becomes scarce, and many of them starve. Population outbreaks of mule deer can also change ecology in unforeseen ways, such as by killing trees when starving deer chew on tree bark. (Large deer populations also assist the spread of Lyme disease to humans.) Populations that are unable to self-regulate depend on predators and other limiting factors for regulation. That is why it is said "wolves are the enemy of deer but the friend of deer populations."

CHANGE IS A CONSTANT

When ecologist Eugene Odum first wrote about ecosystems in the 1930s, he thought that wild populations of plants and animals in an ecosystem would naturally progress through stages of succession toward a harmonious, balanced state. The ultimate stage, such as a mature forest, was called a "climax state."

In later years, Odum refined his ideas on populations, viewing nature as a collection of forces, living and nonliving, that are constantly changing, shifting, and reacting to one another. In this view populations, and even the whole biosphere, never remain static. The balancing mechanisms at work may appear harmonious, but an ecosystem's apparent equilibrium at a given moment will never last. Over time, particularly geological time, change is the only constant.

SPACE AND TIME

When the range of a population shifts, both the ecosystem it enters and the one that it leaves will undergo an adjustment to its ecological balance. For example, one population may dwindle in numbers—even to the point of extinction—when a population of another species takes over its niche and thrives. The abundance and location of one species can cause cascading effects among other, connected species up and down the food chain.

Carrying Capacity

A population that reproduces without limits will grow exponentially. That is, the population will grow faster and faster as each generation multiplies—two children will produce four grandchildren, eight great-grandchildren, and so on. Without limits, it is estimated that a pair of rats could have 20 million descendants in three years.

It is a good thing that in the real world there are always limits. Earth will never be overrun with mice, ivy, spiders, or mushrooms, because each will eventually run up against challenges to its basic needs that will limit its population. When a population grows until it gets as large as its habitat can accommodate, that habitat is said to have reached its carrying capacity. Carrying capacity is the number of individuals that the local resources can sustain.

Lake trout populations self-regulate, producing fewer and slower-growing fish, when food is scarce or when there are too many fish packed into a small space.

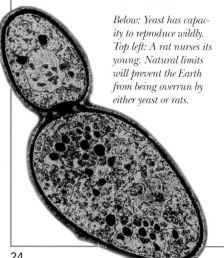

Below: Yeast has capacity to reproduce wildly. Top left: A rat nurses its young. Natural limits will prevent the Earth from being overrun by either yeast or rats.

OVERSHOOTING

A population that draws down its resources faster than they can be replenished naturally is heading for a run-in with the carrying capacity of its habitat. For instance, yeast introduced into a vat of grape juice will initially thrive and reproduce wildly. The yeast, however, will eventually consume most of the sugar. Alcohol and carbon dioxide levels will then rise dramatically, creating conditions that no longer allow the yeast to thrive. All the yeast in the vat will then die.

Individuals in a population that has exceeded the carrying capacity of its habitat may have poor health and suffer from malnutrition because of the compromised living conditions. When this happens, the weakest individuals may die, or the population as a whole may become more vulnerable to further environmental stress or disease. Sometimes a large number of individuals in a population die as a result of overshooting the carrying capacity of their habitat. This is known as a die-off.

Poor Knights Islands Marine Reserve, New Zealand. Setting aside zones where favorable conditions are created for egg survival can help restore depleted fish populations.

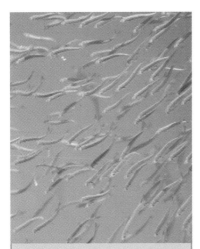

The populations of many ocean fish are declining, but marine reserves can give them a chance to recover.

BUILT-IN SENSORS

Certain animals and plants have a built-in sense of carrying capacity, so that instead of overshooting and having a die-off, they remain within the limits of their habitat's ability to support them. Lake trout, for instance, stop breeding as prolifically when the population density increases too dramatically. Although this is the result of individual responses to chemical signals from other trout rather than a thought-out response on the part of the trout, the result is that population numbers may remain steady for extended periods. The trout will produce more offspring and mature to a reproductive size at a faster rate when populations are threatened, such as when aggressive fishing takes place. When space and food are scarce, such as when a lot of fish are living together in a small pond, the trout remain smaller and reproduce more slowly.

Experiments have shown that no matter what number of lake trout a pond is stocked with in the beginning, the population will increase until it reaches a particular density, then level off at about the same number.

GREEN REVOLUTION

Technological advances in agriculture, called the Green Revolution, extended the Earth's carrying capacity for humans. In Asia, new breeds of rice produced higher yields of edible grain per acre using the same nutrients, so that an acre of rice paddy could support greater numbers of people than before.

A SECOND CHANCE

Exponential population growth is not always a bad thing. Many of the world's fish populations have declined in recent years because of overfishing. Large game fish, such as tuna and swordfish, have been particularly hard hit, but smaller, more opportunistic species have also been decimated. Marine ecologists point out that the remarkable reproductive capacity of some of these species could help restore populations to their previous numbers. For instance, in some species, large females will release millions of eggs each year. Normally, less than one-tenth of 1 percent will survive. If marine reserves were created where large females were protected, breeding conditions were favorable, and fishing prohibited, many of the world's declining fish populations would have a strong chance of recovering.

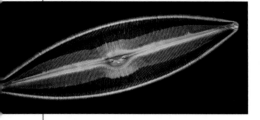

Living Limits

The carrying capacity for a population in its territory is limited by both living (biotic) and nonliving (abiotic) factors. Biotic limits include seasonal fluctuations in food supply, such as when winter sends prey into hibernation deep beneath the snow. Competition may also arise between other individuals in the same species, as when two blue jays compete for the right to control the same territory.

Competition may also develop among several species using the same resources. The massive "dead zone" in the Gulf of Mexico, for instance, is the result of bacteria using up so much

oxygen that there is not enough left for fish, crabs, and other animals. These bacterial population booms result from excess nutrients creating blooms of microscopic algae that feed explosions of zooplankton populations, which creates a lot more dead matter on the bottom to provide habitat for the bacteria.

Sometimes the competition is not only between wild species, but between wildlife and humans, such as when ponderosa pine loggers and white-headed woodpeckers try to make a living from the same tract of forest.

Imagine a population of grizzly bears living in a state park

in Montana. A large part of a grizzly's diet consists of cutthroat trout that swim up the shallow streams to spawn, and the nuts and berries that grow in the surrounding streams and meadows. The number of cutthroat trout living in the shallow streams where they can be caught by the bears, and the size of a season's crop of huckleberries and wild nuts, are both biotic factors that will influence the health and size of the grizzly population.

BARRIERS TO GROWTH
Nonliving, or abiotic, limits include soil quality and water supply. Houseplants and

Above left: Gray wolf on the prowl. Above right: Grizzly bear catching fish in a stream. Predators such as the gray wolf and the grizzly bear are living, or biotic, limits to the populations of their prey. Top left: Although a single diatom alga is microscopic, clusters of algae can trigger a cascade of effects that limits marine life.

Environmental factors such as extended droughts can be barriers to growth. When water is scarce, vegetables and other plants are unable to thrive. This desiccated ear of corn on a plant in a field suffering a drought demonstrates one of the effects of an abiotic limiting factor.

vegetables grow to greater sizes with rich fertilizer and adequate amounts of water. It is the same for wild plants; when wild plants are well fed, they will grow, and when they are deprived of water, sunlight, or nutrients, they will perish.

Other types of abiotic limiting factors include geographic barriers, such as mountains or rivers that prevent migration. Human settlement is an increasingly common geographic barrier for many species.

In an extended drought, the populations of many plants will be limited by the lack of water. Animal populations as well are affected by drought. Not only is food more scarce, but the need for water may drive animals into situations where they are more vulnerable to predators. And

many animals, like frogs and fairy shrimp, depend on the presence of ponds for reproduction.

The population of bears mentioned above, as well as the population of huckleberry bushes, will have a better chance of surviving and expanding if the bears can migrate to new ranges when drought diminishes the food supply. If they are confined to a small range by park boundaries, the biggest limiting factor on their population will become humans.

MANAGING POPULATIONS

In the winter, farmers often plant cover crops, such as nitrogen-fixing clover, to increase the soil's ability to support healthy populations of crop plants in season. Ranchers of grass-fed cattle regularly move their herds from pasture

to pasture so that the grass and clover that feed the cattle in any single field will not become overgrazed. In artificially bounded wild settings such as national or state parks, wildlife managers act systematically to limit plant and animal populations. Certain trees and shrubs may be planted to attract and support birds, butterflies, and rodents.

Animals such as deer, whose population explodes in the absence of natural predators, may be "culled," or killed off, so that they do not overgraze the vegetation. Predators may be reintroduced into the landscape, as wolves have been in Yellowstone National Park. Understanding the natural carrying capacity and limits of the land is important to designing good management strategies.

CHAPTER 3

PARTNERS IN EVOLUTION

Left: The southern yellow-billed hornbill is common to parts of Africa. It prefers seeds and small insects, but in the dry season will eat termites and ants. Hornbills often lay their eggs in the trunks of baobab trees. Top: The elephant is among Africa's native population of large mammals. Bottom: The baobab tree grows in low-lying areas of Africa. It can create its own ecosystem, supporting the life of creatures from birds that nest in its branches to elephants, which have been known to knock down and consume these trees.

Communities of organisms have evolved many ways of sharing the living spaces available to them. As long as individual species do not compete too heavily for sources of food and shelter, dozens of species can live together harmoniously on one tree, or one stretch of shoreline.

The interconnection of feeding relationships among multiple creatures—a predator and its many prey items, an organism and all those trying to eat it—is called a food web. The complexity of interactions means that what affects one species may have unanticipated results elsewhere. Predators who limit one species of prey, for example, may indirectly benefit another species. Sometimes the interactions are so intimate that two creatures effectively become one, such as bacteria that live in an animal's digestive system. Neither the bacteria nor its host would thrive without the other.

An animal with a significant effect on its ecosystem is called an ecosystem engineer. Invasive species can also have an oversized effect on an ecosystem, since they are not part of the food web that has evolved in that location and consequently are able to dominate the system because they have no natural predators in the new location.

Keystone Species and Ecosystem Engineers

Kelp forests nourish and protect many kinds of animals, including sea otters.

Some species seem to have a stronger influence than others on their ecosystem. Take away the ocher sea star along the Northwest coast of the United States, for instance, and the ecosystem changes dramatically; in the absence of these sea stars, their favorite prey, mussels, takes over and makes it hard for other species that used to live there. Sea stars are known as keystone species, because as top predators they determine ecosystem structure by their eating habits.

If you chop down an aspen tree by a beaver pond, not much will happen; but if you take away a beaver, a wetland might dry out, changing the kinds of plants that live there and the animals that rely on them. Because beavers exert their influence by physically altering the landscape, they are known as ecosystem engineers. Even minute organisms can be ecosystem engineers. The massive calcium carbonate structures built by tiny corals radically alter the ecosystem around them, protecting the shoreline

Top left: A beaver dam. Right: Sea otters can be found in the coastal waters of the northern Pacific Ocean.

and creating a complex habitat in which numerous fish and invertebrate species can live.

SEA OTTERS
Kelp forests off California are so rich in diverse life that they have been called rain forests of the sea. Hundreds of species, from bonito to jellyfish to grebes, depend on these fast-growing sea plants—giant kelp are the world's largest algae and grow up to 200 feet tall—for food, shelter, or both.

One of those species is the sea otter. The whiskered otter basks on her back like a sunbather at the beach, often secured to a piece of kelp so she will not drift away, as she cracks open snacks of abalone or sea urchin on her stomach. Sea otters are a keystone species of Pacific coastal waters. An otter eats as many as 50 large sea urchins each day. Its feeding limits the population of urchins, which eat giant kelp. Without the balancing presence of sea otters to keep urchin populations down, kelp forests disappear, and with them goes a habitat for fish, worms, abalone, and dozens of other marine species.

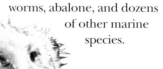

After sea otters were hunted to near extinction in the nineteenth century, kelp forests off the Canadian and U.S. Pacific coast suffered major declines. Kelp forests off California decreased by more than 80 percent. With the help of conservation efforts, sea otter populations have recovered to several thousand individuals; nonetheless, wildlife managers are still having a difficult time restoring the kelp forests.

BEAVER ENGINEERS
Beavers are another highly influential species, shaping ecosystems and enabling other species to thrive by engineering water systems. Beaver dams turn streams into wetlands, ponds into lakes. The productivity and biodiversity of the beavers' environment rises because of the increased moisture.

Wetlands created by beaver dams are often bordered by denser vegetation than surrounding areas. In dry regions, these streamside, or riparian, landscapes support trees and shrubs that shelter migrating birds and resident animals. The roots of plants at water's edge are dense and deep, controlling erosion and holding moisture in the soil.

Like sea otters, beavers were once intensively hunted for their silky fur. Collecting beaver pelts for the top hat trade was one of the main economic reasons for the opening of the western frontiers of North America, beginning in Canada as early as the late sixteenth century.

At the same time that the beaver population was steadily shrinking, the interior of the North American continent was being carved into farms, and humans, with their man-made irrigation systems, became the ecosystem engineers.

KEYSTONE CONSERVATION

Because of their critical role in shaping ecosystems, keystone species and ecosystem engineers have become a major factor in conservation planning. In smaller African reserves, for example, elephant herds are culled to keep them from having too big an influence on their now-limited ecosystem. Black-tailed prairie dogs in the American interior are hated by farmers but beloved by prairie restorationists because without them many plants and animals will not be able to survive. Nine species, including

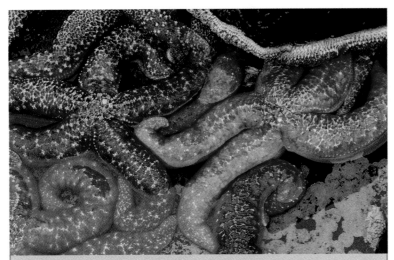

Starfish are found in intertidal areas where they hold fast to rocks with thousands of tiny tube feet, many of which have suction cups on the ends.

STARFISH PROTECT DIVERSITY

In a classic ecosystem experiment, ecologist Robert Paine discovered that the starfish *Pisaster ochraceus* played a keystone role in Washington State's intertidal zones. Fifteen species of shellfish and barnacles coexisted on the Pacific rocks and were prey for the starfish. But when Paine removed the starfish, the number of species remaining dropped to eight.

Paine discovered that in the absence of the starfish, mussels crowded out the limpets, chitons, periwinkles, and other mollusks that shared their habitat. The top predator starfish, far from limiting the populations of the mollusks by eating them, actually protected the weaker prey. The starfish kept diversity in the intertidal zone. Paine said of his research, "It's importance is that it convinced managers and conservationists alike that the ecological impact of single species matters."

black-footed ferrets and burrowing owls, depend on the prairie dog for both food and housing. Dozens of other animals, birds, and plants eat prairie dogs, live in their burrows, or benefit from the soil aeration, grass-cropping, and other things prairie dogs do to their environment. Nonnative ecosystem engineers, such as cordgrass on the West Coast of North America, are seen as particular threats.

The damage caused by foraging prairie dogs to American farms has led to extensive eradication programs. Conservation experts argue that prairie dogs may actually be beneficial to the land's fertility.

Close Relations

Creatures evolving in the same place sometimes develop close relationships. One might masquerade as another, or provide food for another, such as the cleaner wrasse living on a coral reef whose diet consists of other fish's parasites.

Cattle egrets supplement their diet by picking insects off cattle, whether Brahman bulls in India, water buffalo in China's Sichuan province, or wildebeest in Kenya; they have even been seen with kangaroos in Australia. Acacia trees have evolved a close relationship with acacia ants, which are aggressive creatures that swarm herbivores that try to eat their host tree.

A relationship between two organisms may be beneficial for both parties, or bad or neutral for one but good for the other. Vines use their ability to climb trees to get closer to the sun, their energy source; yet often a vine covers so much of its host tree that the tree withers. The cholera virus, as a much smaller example, needs its host in order to replicate and spread, but the host gets nothing but illness from the transaction.

INSECT MASQUERADE

The drone fly looks just like a bee, and the hornet fly resembles a hornet. Neither fly has a stinger, but both use their beelike coloring to fool predators into thinking they do and leaving them alone.

Other animals avoid predators or ambush prey by looking like a thorn, a green or dead leaf, lichen, bark, or like poisonous insects. The viceroy butterfly looks a lot like a monarch. Since the monarch butterfly is poisonous to eat, birds mistaking the viceroy for a monarch will not eat him.

Above: A cattle egret rides on the back of a water buffalo. These birds hold on with sharp claws and feed on skin parasites such as ticks. Top left: An acacia tree in Dubai. Ants protect acacia trees from invading insects while the tree provides shelter for the ants.

A leaf insect, also known as a walking leaf, camouflages itself with large fore-wings that resemble withering green leaves.

A stick insect is nearly indistinguishable when it sits on a branch. Other creatures' camouflage is less thorough but still effective. The chameleon and the pepper moth, for instance, change their appearance to match their surroundings. Some insects, such as grasshoppers, also use behavior as camouflage, for instance by swaying like a leaf in the wind.

PARASITES

Parasites are organisms that get their nutrients from another organism without killing the other organism outright. Ticks that live on deer blood are parasites; although not generally harmful for the deer, some of these ticks can spread diseases to humans. Microbes in deer ticks that cause

Lyme disease in humans are also parasites. Ticks are the vector for the disease: Vectors are the transportation taken by the bacteria to get to their next host, in this case humans.

DIGESTIVE HELPERS

The bacteria living in human intestines help break down food into nutrients that human bodies can burn for energy. This process helps us use food more efficiently, so that there is less unused material to be emitted as solid waste and methane gas. Digestion of food with help from bacteria is one example of how two organisms can coexist peacefully for each other's benefit.

Similarly, the fungus residing in an African termite mound exists in a state of mutual benefit with its termite hosts. Fungus-farming termites in Africa cultivate a single strain of fungus in moist chambers within their mounds. The termites feed the

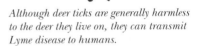

Although deer ticks are generally harmless to the deer they live on, they can transmit Lyme disease to humans.

fungus chewed wood and grass pulp that would otherwise be indigestible, and the fungus breaks the pulp down and converts it into nutrient bits that the termites can use.

Other species of termites have evolved gut bacteria that help extract nutrients from chewed raw materials, in a manner similar to humans. Most plants have symbiotic fungi living on their roots, helping them to absorb needed nutrients, such as phosphorous and nitrogen, from the soil.

Termites feed primarily on wood that contains a high level of cellulose, although they cannot digest the cellulose themselves. They rely on microbes in their guts to do the digesting for them.

Niche Living

Most creatures organize their lives around a niche, a physical territory and food source that no one else in that ecosystem has claimed. This is a way of ensuring survival by cutting down on competition with other creatures for food, water, sunlight, shelter, or other essentials.

A woodpecker living off bugs under the bark of a white oak does not compete with a squirrel that lives in the hollow of the same oak and eats its acorns. A potato bug in a garden can live in harmony with a tomato hornworm the next row over and across the path from a cabbage moth. Beetles are such a numerous and successful group of species partly because they develop in two stages. A wormlike baby beetle larva has its own niche and its own food sources, sometimes in water or soil, that are separate from the niche and food of its parents, and those of its own mature form. By occupying two separate niches, one in youth and one in adulthood, beetles reduce the competition for resources between parents and children.

BODY LANGUAGE

Over time, some organisms evolve special talents or special body parts that make them better adapted to a particular niche. Anteaters, for example, have developed extra-strong front claws and a long, thin tongue that allow them to get at succulent ants and termites inside concretelike clay mounds. Charles Darwin himself discovered an orchid in Madagascar that had an enormously long spur in back of its stamen, and which was fragrant only at night. Darwin predicted that a moth would be discovered with an enormously long proboscis that was adapted specifically for this orchid. Fifty years later scientists found Darwin's hawk moth, whose proboscis is an incredible 12 to 14 inches (30–35 cm) long.

Left: Woodpeckers can create fist-sized holes in trees. They use their long, sticky tongues to reach into the holes and retrieve carpenter ants and other insects. Right: Ladybug larva eating aphids. Top left: A squirrel on an oak tree.

Two different species can develop tightly linked lifestyles, and changes in one will necessitate changes in the other. Coevolution is a term for this interdependent evolution. The selective forces may be negative (predator and prey) or positive (pollinator and flowers). A well-known example of coevolution is the relationship between honeybees and flowers. Flowering plants need some way to move pollen from one flower to another in order to reproduce. Bees travel from flower to flower collecting pollen to make honey for food, but as they do so they carry pollen as surely as if they were the flowers' servants. The bees' pollination work for the flowers also ensures a greater supply of honey for themselves.

The work of a bee benefits both the bee and the flower. The bee carries pollen back to the hive to make honey and from flower to flower, helping the flowers to reproduce.

WORLD ON A BAOBAB

The hardy African baobab tree serves as a habitat for dozens of animals, each occupying a different niche in the giant tree. Bush babies, tree frogs, snakes, lizards, kestrels, parrots, scorpions, spiders, insects, monkeys, and squirrels all live on the same tree at once.

A hornbill lays her eggs in one cavity in the huge tree's trunk, an owl in another, while a stork makes a nest from sticks on an outer branch. The baobab's flowers are pollinated by fruit bats and bush babies. Elephants and baboons eat the fruit. Giraffes nibble on the leaves. Geckos hide in the folds and crevices of the spongy wood. Pools of water collected in depressions where the great branches come together become reservoirs for animals as surely as watering holes on the savanna below. The ability of an ancient baobab to provide niches for so many creatures has given it a reputation as the original "tree of life."

With a lifespan of up to 3,000 years, a single baobab can provide shelter for many different animals, including rodents, lizards, and spiders.

Native Species

As we have seen, the numbers of creatures in an ecosystem are limited by predators, food supplies, and places to live, as well as by competition and reproductive ability. In ecosystems where the insects, plants, and animals have evolved together over time, these forces keep populations in balance so that the overall diversity of the ecosystem is maintained. Sometimes, however, an organism travels to a new place where its natural checks and balances do not exist. This happens evermore frequently in a world where vegetables and tourists alike travel across the globe in a single day. Although some newcomer organisms die off quickly, others successfully invade the new territory and compete with the native life. In these cases, the non-native is called an invasive species.

Invasives can be plants, animals, or even viruses, bacteria, and other single-celled organisms. Brown tree snakes have so far eliminated 13 species of birds on the island of Guam; a few of these species have become extinct. Asian longhorn beetles are threatening maple trees in New England. Starlings from Europe have made bluebirds in the eastern United States quite rare. According to The Nature Conservancy, invasive species are one of the biggest threats to biodiversity on the planet, second only to habitat loss.

CROWDED OUT

Water hyacinths are choking waterways from Australia to Zimbabwe. The dense, floating weeds, which have a showy, purple-blue flower, can take over physically, weaving thick mats that are difficult for both boats and swimming animals to penetrate. The mats also take up nutrients that other plants need to survive, and block the sunlight.

Lakes and rivers choked by water hyacinths eventually become unusable to other plants, algae, fish, and animals. Some fish may disappear entirely when their habitats are taken over by water hyacinths, making survival difficult for animals that depend on fish as a primary

Above: Water hyacinths choke waterways, making them unusable to other plants and wildlife. Top left: Threatened native reeds.

Certain species of mussels are invasive; these non-natives compete with and crowd out those organisms native to the area.

food source. In Florida, the snail kite is endangered because its favorite snails have become difficult to find; this is because the snails' favorite plants have been crowded out by water hyacinths.

Zebra mussels have similarly crowded out native shellfish in waterways where they have been introduced. After being dumped with ships' ballast in the Great Lakes, the mussels have been so successful that they are not only threatening the life of native shellfish but also wreaking billions of dollars in damage to water-treatment and other waterside facilities.

FIGHTING INVASIVES
People can help fight and control the spread of invasive species by planting only native or noninvasive plants in their gardens and backyards. Aquarium pets should never be released in lakes and streams or flushed down the toilet.

In the case of invasive water hyacinths, scientists are working on introducing the hyacinth's natural pests into the plants' new environments. There are several

leafhoppers, flies, and other insect species that harm or kill water hyacinth in its native Amazon River basin. Fighting hyacinths with such insect pests currently seems the best hope for controlling the species. The danger is that the pests, once moved from their native habitat into a new environment, will cause more problems for the new ecosystems than the water hyacinths do.

WORMS CHANGE FOREST
Even invasive species that do not seem dangerous at first can negatively impact their new environments. Fifteen species of earthworms that have been introduced into the Minnesota woods are threatening the ecology of the hardwood forests there. The worms arrived in the soil of imported plants and were released in other areas as unused bait.

If there were native earthworms in Minnesota, they did not survive the last ice age. After the retreat of the glaciers, the woods adapted to slower decomposition without worms; the plants and animals became accustomed to a thick layer of dead leaves, called duff. As the imported earthworms spread, they eat the leaves and decompose the duff faster than normal, taking away habitat from organisms adapted to living in the duff and changing the way water runs through after a rain.

Since there is no practical way to remove worms once they take hold, ecologists are trying to make sure that areas that are still worm-free remain so.

The native plants that surround a pond will soon be crowded out by invasive purple loosestrife.

CONSERVING NATIVES
Native plant societies in many locations encourage people to cultivate local plant species, rather than the exotics that are sold in many garden centers. Sometimes conserving native plants means conserving a whole landscape, such as a valley, where rare species still thrive. Native plants often have benefits for wildlife as well as conserving the distinctive look of local landscapes. In Florida, the native plant society encouraged nurseries to stop selling dozens of non-native plants. Some of these were banned by the Department of Environmental Protection because of their potential harm to local ecosystems.

Trophic Cascades

The creatures in a food chain often can be diagrammed in a single line; it is easy to draw a link from one to another. An orca eats a sea lion, a sea lion eats a salmon, a salmon eats a fly, a fly eats algae, and algae are primary producers using the Sun's energy to make food.

In food webs, however, the lines are more complex, since most predators have more than one kind of prey, and many different predators will eat the same species of prey. Sometimes a meat eater can affect the plants in its ecosystem by what is called a trophic cascade. For example, wolves eat elk, and elk eat willows. When wolves were hunted out of Yellowstone National Park, the elk population began to rise, and because there were greater numbers of elk eating streamside plants, these plants decreased in number.

Since wolves were reintroduced into Yellowstone, the streamside landscapes, including willows, have been restored, because the population of the plants' worst enemy, the elks, was being limited once again by the wolves.

TOP UP, BOTTOM DOWN

If the population of creatures in a food web is mostly limited by predators, it is said that the population is controlled from

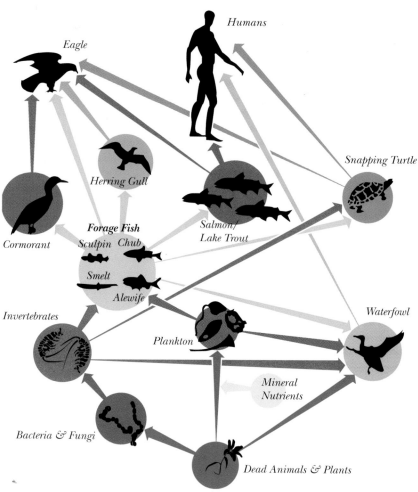

Above: A complex food web. Top left: Orca whales are at the top of the ocean's food chain.

the top down. If the population becomes limited by resource constraints (problems of not finding enough food, water, fertilizer, mates, or other necessities) then the population control is bottom up. By understanding the degree to which populations are being

controlled by bottom-up or top-down forces, ecologists can more prudently manage ecosystems.

Ecologists can study a food web by looking at the interactions among creatures large and small that connect through the web. Another way of looking at

An elk standing in a stream at Yellowstone National Park. When the number of wolves decreased in Yellowstone Park, the number of elk increased significantly. The increase in the elk population meant more elk nibbling on streamside plants, which in turn lowered the number of these plants, upsetting nature's delicate balance.

a food web is by tracing the flow of energy, water, biomass, and nutrients through the system.

CONCENTRATING POISONS

As each predator in a food chain eats a creature lower down the chain, toxic substances in the lower creatures can become concentrated in the body of the higher predator. Levels of certain contaminants are frequently thousands of times higher in top predators than in the environment in which those predators live. This biomagnification, as it is called, almost wiped out bald eagles, and strongly affected

populations of other birds of prey. In this case the contaminant was DDT, once widely used as a pesticide. When these birds ate fish, they also ingested DDT that the fish had taken in from algae and from the water. The effect of accumulated DDT on the top predators was to make the walls of their eggs too thin for their chicks to survive.

In 1962, ecologist Rachel Carson (1907–64) wrote *Silent Spring*, which documented how pesticides like DDT could have a major effect on top predators. After *Silent Spring*, and the work of some environmental

organizations, the United States banned the use of DDT. Without DDT in the food chain, the population of eagles and other hawks began to rise once again.

The bald eagle population was put at risk by eating fish contaminated with the pesticide DDT, which was banned in the United States in 1972.

CHAPTER 4

A SENSE OF PLACE

Left: An elf owl (Micrathene whitneyi) *peers from a saguaro cactus nest cavity in Sabino Canyon, Tucson, Arizona. Top: A water strider on the surface of a stream. Bottom: A golf course in Las Vegas, Nevada. Such human water uses may affect the survival of wildlife nearby.*

The morning sun slowly slides over a giant saguaro cactus, and a cast of characters uniquely adapted to the desert ecosystem—owls, snakes, kangaroo rats, scorpions—goes to sleep for the day. This community of species, along with the rest of its biological community of plants, other animals, and microscopic bacteria, and its nonliving environment of matter and energy, is known as an ecosystem.

An ecosystem can be as big as a biome, a region such as a deciduous forest, defined by its family of distinctive life forms, climate, geography, and elevation. An ecosystem can also be as small as a pond, with its bullfrogs and cattails, algae, minnows, and dragonflies. Like a machine whose moving parts function as a larger whole, the pieces of an ecosystem affect one another, fueled by energy and nutrients cycling through their parts.

How does a population of mosquitoes affect striped bass and egrets in a marsh? How much water can be taken from a desert by farmers and cities while leaving enough for wild species to thrive? Although ecosystems have no sharp boundaries in the real world, they are a convenient concept for ecologists to use in understanding a place.

Biomes

Large ecological regions on land that share similar climate and vegetation are called biomes. Major types of biomes include deserts, grassland, forests, mountains, chaparral, and tundra. These categories can be further broken down into more specific biomes, such as coniferous, deciduous, and temperate or tropical rain forests.

Characteristics often shared by biomes in different parts of the world include temperature and rainfall. These are two of the most important aspects of climate. Life that develops under similar conditions in two different places on the globe may be similar even though it does not actually share a genetic ancestry.

A map of the world's biomes shows borders that cross national boundaries. The northern, or boreal, forest, for example, extends in a belt all around the globe, spanning Canada, northern Europe, and Russia. Following are some simple explanations of a few of the world's biomes.

TUNDRA

North of the boreal forest, the Arctic tundra is a distinctive biome in which plants are small and tough, and lichen—which is a community of bacteria and fungi living symbiotically—is a primary food source. Plants cannot afford to have leaves with large surface area, as these would freeze. Since very little sunlight is

The giant saguaro cactus thrives in the arid climate of the Sonora Desert in southern Arizona and northwestern Mexico.

Above: The Innoko National Wildlife Refuge in Alaska is part of the boreal forest biome that extends in a global belt, across several continents. Top left: The towering peaks of Grand Teton National Park in western Wyoming are part of an alpine, or mountain, biome.

available for photosynthesis during winter months, tundra plants must have the ability to become dormant for long periods. Both plant and animal life must be adapted to endure long, cold winters with little food.

DESERT

Deserts are defined as places where rainfall is less than two inches (50 mm) per year, or where evaporation is greater than precipitation. Some deserts are hot and dry, such as the Sahara, with very little life except around groundwater-fed oases. Others are arid lands but with a fairly large amount of life, such as the Sonora Desert of Arizona and northern Mexico with its famous saguaro cactuses.

Plants and animals in deserts cannot depend on regular supplies of water and so must store or conserve water. Succulent plants, of which cactuses are one family, are adapted to store water in their fleshy trunks. Net primary productivity, the amount of biomass or energy stored by plants to provide fuel for all life in the ecosystem, is fairly low in desert biomes.

TROPICAL RAIN FOREST

In their warm and rainy biome, tropical rain forest plants grow very large leaves to help disperse heat and water. Nutrients are virtually all stored in the plant mass; decomposition and nutrient recycling happen rapidly, and the soil is very poor. Microclimates at various heights, from understory to canopy, allow separate ecosystems to live on each level. Rain forests rank high among the world's most productive ecosystems, and is also a biome containing some of the world's greatest concentrations of biological diversity.

DECIDUOUS FOREST

In places with warm summers and cold winters, many trees drop their leaves in winter and become relatively dormant, growing new leaves again when days grow longer and weather warms up in the spring.

Animal life is adapted to four seasons, hibernating, migrating, or storing food supplies for winter and figuring out how to stay sufficiently warm in the cold months. Because the soil in deciduous forests contains a lot of stored nutrients, these forests have often been converted to agricultural land by humans.

GLOBAL VIEW

Because biomes help scientists to make connections between one place and another, they encourage a global perspective of climate and life. Viewing ecology through a biome perspective can give scientists another prism through which they can understand how global changes, such as climate change, are affecting life all over the planet.

Evidence of climate change, in fact, rests partly on observations of the distribution of species in various biomes. Evidence of climate change is strengthened when ranges are observed to be shifting in ways that support the effects predicted by climate change scenarios.

The Montana prairie is part of the Great Plains, a temperate grassland in the middle west of North America.

SIMILAR GRASSLAND

The temperate grasslands of Mongolia (called steppe) are strikingly similar to those of Argentina (called pampa) and the native grasslands that once stretched from horizon to horizon in the center of the North American continent (prairie). Why have such similar landscapes developed on three separate continents? Rainfall or water availability is an important determinant of what types of plants thrive in a particular location. Elevation, soil type, and average temperatures also affect what types of plants will grow in one place.

Microclimates

A wooded valley with a stream running through it is cooler, shadier, and more humid than the meadow just next door. In the meadow the morning dew will evaporate in the sun. As a result asters and goldenrod will thrive in the meadow, while fern and trout lily will stay where conditions are moist and shady.

Meanwhile, in the Northern Hemisphere the north-facing side of a tree will grow moss, while the other side's bark is relatively free of moss, because the north side is cooler, shadier, and moister than the other. The bark and the moss, insects that live in them, and the woodpecker that visits the tree to dig them out, coexist in a little ecosystem.

These two examples of small ecosystems depend on microclimates, places where geography or local plant and animal life make climate conditions special in one small area, different from the general surrounding climate.

COASTAL ECOSYSTEMS

Coastlines are a place where several ecosystems come together. The area between high and low tides, called the intertidal zone, is home to organisms such as fiddler crabs, barnacles, and various algae that can withstand being alternately wet and dry, hot and cold, as tides rise and

Above: The remoteness of Washington State's Olympic Coast spares intertidal communities from the pressure of sheer numbers of people. In this wild environment, habitats remain intact and organisms interact through natural processes, including competition for space and predation. Top left: A shaded patch of creek can harbor its own microclimate.

Above: An early photo of the Lincoln Memorial: Cities, towns, and farms worldwide have destroyed large swaths of wetlands. Top left: Water is released from a factory. Human uses of natural resources are not without ultimate costs. Top right: Collecting water samples for acid rain analysis in a Chesapeake Bay wetland tributary.

FRESHWATER SYSTEMS

Freshwater lakes and streams are an important type of ecosystem because they support life along a large geographical space that crosses many borders. Lakes and streams and their shorelines abound with life, from reeds to stone flies to humans, and the same lake or stream may be used by all of these organisms.

When a freshwater system is healthy, its water contains plentiful amounts of dissolved oxygen, as well as plankton, insects, snails, and fish. Organisms in the riverbed or lake bottom are alive and well. The system does not contain levels of chemicals that make the water unhealthy for the creatures that live in it.

Scientists who manage freshwater systems keep track of the water quality by sampling and analyzing it regularly. They try to create and maintain conditions that work for wildlife and humans. Sometimes there is conflict. A real estate developer might want to drain a wetland to build a housing development (for example, large parts of Washington, D.C., and many other cites worldwide are built on former swampland), but the wetland might be important both as a nursery for fish and for flood control.

Deciding which side to value over the other—flood control or housing, baby fish or real estate developers— is the job of policy makers, who include both elected officials and ecologists.

fall. Shorebirds such as egrets, herons, oystercatchers, and willets find a feast when the tide is low. Many birds nest in the sand above the high-tide line, or in nearby trees, and they also depend on the shoreline to guide them on their seasonal migrations.

Worldwide, half of all people live within 100 miles (62 km) of a coast, either of the ocean or a major freshwater body. Most large cities, from Shanghai to New York to Rio de Janeiro to Athens, are located along coastlines.

People enjoying the ocean beach in summer flock to the shore to soak up the sun and swim in the waves. In many parts of the world, fish caught along the coasts are an important supplier of the protein in the population's diet. But this human population's factory effluents, sewage, and agricultural runoff may eventually pass through the same intertidal zones that are home to the fiddler crabs, baby striped bass, and razor clams.

Nutrient Cycles

The atoms and molecules needed to support life flow through Earth's ecosystems in continuous cycles. The law of conservation of matter, described in the eighteenth century by French chemist Antoine Lavoisier, implies that these nutrients can be tracked as they transform from plants to animal bodies to soil, and so on. With the exception of energy provided by sunlight and the occasional meteor or meteorite, the Earth is essentially a closed system.

The nutrient cycles that support life on Earth include nitrogen, phosphorous, and carbon. All of these elements cycle through air, ground, and water, and through the living bodies of Earth's creatures and the fabric of the soil.

THE NITROGEN CYCLE

The element nitrogen is an important ingredient in the organic compounds that make up the building blocks of life. These compounds include amino acids, proteins, DNA, and RNA.

In general, in coastal and terrestrial ecosystems, usable nitrogen is in short supply and limits the populations of both plants and animals. Almost 80 percent of Earth's atmosphere is nitrogen gas, but in a form that cannot be used as a nutrient by living organisms.

NITROGEN FIXERS

Fortunately, nitrogen in the atmosphere can be converted into compounds that can enter food webs. This conversion process, called nitrogen fixation, is performed by specialized bacteria.

Most nitrogen fixation occurs through bacteria living in symbiosis with a family of plants called legumes. The large legume family includes types of trees, vines, and shrubs. Some well-known legume family members are alfalfa, clover, beans, peas, lupines, peanuts, and lentils.

The nitrogen-fixing bacteria, called rhizobia, live in nodules on the roots of plants in the legume family, where they convert atmospheric nitrogen to the kind of nitrogen organisms can use. If you pull up a bean plant, you can observe these

Above: Legumes such as peas (pictured here), beans, and alfalfa are essential link in Earth's nitrogen cycle. Left: Peanuts, a versatile member of the legume family. Top left: Lupines are a member of the large legume family.

Nitrogen runoff from manure and other fertilizers used on farms in the American Midwest has been blamed for a "dead zone" off the coast of Louisiana.

little nodules, which turn from white-green to pink as nitrogen is fixed.

When the bacteria die and the plants decay, nitrogen is released into the soil in a form that can be taken up and used by other plants and the animals that eat them.

EUTROPHICATION

Paradoxically, too much nitrogen can present problems for an ecosystem. Nitrogen-rich runoff from manure, other agricultural fertilizers, and untreated garbage and sewage, for example, has been linked to oxygen-starved water systems and harmful algae blooms throughout the world. Eutrophication, or overenrichment of a water body with nutrients, results in an excessive growth of algae and oxygen starvation that kills off other organisms.

In the United States, nitrogen runoff from dairy, poultry, and corn farms in Iowa and Illinois is partly to blame for a dead zone off the coast of Louisiana larger than the state of Massachusetts, as well as blooms of toxic algae that can be harmful to wildlife and humans.

*Nitrogen-fixing nodules on the roots of a pea plant (*Pisum sativum*) allow the plant to utilize free nitrogen in the atmosphere and soil. Symbiotic bacteria in the nodules fix the nitrogen, transforming it into a form usable to the plant.*

TREE FALLOWS

The corn crops growing in villages near Kampala, Uganda, are meager. Local farmers and their families often go hungry because they have not produced enough corn to sustain themselves until the next crop can be harvested. Their soil contains too little nitrogen and phosphorous, and they cannot afford to buy imported fertilizer.

Recently, Kampalan farmers have begun to increase their soil's nitrogen content by planting leguminous trees between the corn rows. These trees fix nitrogen in the soil and also hold in moisture. Just by changing their planting techniques, farmers can increase their soil's nitrogen content and improve their corn crops enough to eat corn throughout the year—and even have some left over to sell.

In Uganda, right next to farmers struggling with nitrogen-poor soil, excessive nitrogen and phosphorous in Lake Victoria flows from untreated city garbage and sewage, as well as from expensive imported fertilizer used on export crops such as flowers. The resulting eutrophication, in combination with problems caused by a non-native Nile perch, threatens the health of the 30 million people who depend on the lake for water and food.

FERTILIZER AND FUEL

Man-made inorganic fertilizer, invented during World War II, has added to the nitrogen released by humans into the nitrogen cycle. Nitrous oxide created when we burn fossil fuels also increases the amount of nitrogen that can be used by living things. These sources have helped increase agricultural yields, but they also contribute to environmental problems such as acid rain and dead zones in lakes and oceans.

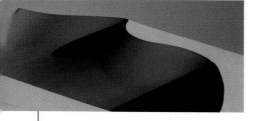

Water in Our World

All living things depend on water. Though water covers nearly three-quarters of Earth's surface, it is distributed unevenly. Some places have plenty of water, while others live in a constant state of scarcity; a single location may have too much water some of the year, and not enough at other times. The amount of water in an ecosystem has a strong effect on how many organisms can survive and grow there. Life of all kinds is much rarer in very dry climates.

When too much water is used by one member of the community, other organisms may suffer. Weeds siphon water needed by food crops, and humans watering desert golf courses may affect the survival of wildlife downstream.

THE WATER CYCLE

Marine ecologist Elliot Norse observed that in every glass of water we drink, some of the water has already passed through fishes, trees, bacteria, worms in the soil, and many other organisms, including humans.

In a cycle that never ends, water evaporates from the Earth's lakes and oceans and then moves through the atmosphere until it condenses and falls as rain or snow, then runs through rivers and percolates through soil until it returns to storage in lakes and oceans.

Plants affect the water cycle by reducing evaporation from the soil and by releasing it slowly through leaves in a process called transpiration.

Waterwheels known as noria were built by the Romans in northern Africa as early as 1 BCE. Surviving examples of noria include the world's largest, at nearly 90 feet in diameter, and the oldest, from approximately 1,000 years ago.

Transpiration is why heavily forested areas have more humidity in the air than areas with less plant life; clear-cutting actually reduces humidity and rainfall. As water runs through the land, it nourishes plants and provides drinking water and habitat to animals. Water also carries nutrients between ecosystems; the Egyptians depended on annual floods to deposit silt on the soils of the Nile Valley to fertilize the year's crops.

SPECIAL CHEMISTRY

Water's special chemical properties create conditions for life to thrive. For instance, the strong bonds between water's hydrogen and oxygen molecules enable water to absorb heat as it evaporates and release heat as it condenses,

Top Left: Shifting desert dunes. In the absence of water, life becomes rarer. Above: A jungle waterfall. Abundant water is key to the abundant life in a tropical rain forest.

PRECIPITATION

CONDENSATION

EVAPORATION

RUNOFF

TRANSPIRATION

INFILTRATION

Six components make up the water cycle. Transpiration releases water through leaves. In evaporation, surface water gets absorbed into the atmosphere. In condensation, water in the atmosphere forms clouds. When clouds cannot hold any more water, precipitation brings the water down to Earth, where it either goes directly into the soil, in a process called infiltration, or flows into bodies of water, as runoff.

and to gain and lose significant amounts of heat without changing form; this allows water to distribute heat around the planet.

TIMING AND LOCATION

The timing and location of rainfall are major factors affecting plant and animal populations. In many areas, the arrival of spring, with warmer temperatures and an increase in water supply, is a cue to species from maples to trout that it is time to reproduce.

Ecosystems around the world have become finely tuned to their local water supply. In the Pacific Northwest, for example, some salmon migrate upstream to spawn just as winter snowpack melts and increases the amount of water flowing through streams. At the same time the increased moisture provides breeding grounds

for flies and gnats that are the breeding salmon's food supply.

CLIMATE CHANGE

If the amount or timing of water availability changes, the plants and animals that have come to depend on the usual water cycle may suffer. The global average surface temperature has been growing warmer. A warming climate speeds up evaporation and lets the air hold more water

vapor before condensing. Fewer but more extreme rainstorms can result, which cause flooding and erosion and leave dry spells in between, making life difficult for both plants and animals.

Warming climates' threats to the water supply are particularly troubling for regions that are prone to drought and for those that depend on snowmelt for the spring growing and breeding season.

A warming climate will bring changes in global water distribution. One expected result is more extreme rainstorms, with consequent flooding.

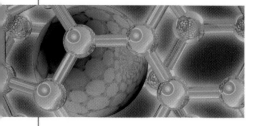

Carbon Basics

Carbon is the element that is common to both coal and diamonds. It is a basic ingredient of life, essential to DNA, carbohydrates, and proteins. The bonds between carbon atoms, and between carbon and other molecules, store most of the chemical energy needed for life.

Carbon dioxide (CO_2) is one of the "greenhouse gases" that help to regulate Earth's temperature and thus are essential to life as we know it. And although CO_2 is only a small part of the atmosphere, increased concentrations of it are a significant contributor to global climate change.

THE CARBON CYCLE

Carbon circulates in the atmosphere and in the ocean mostly as carbon dioxide. Plants, the base of most terrestrial food chains, use photosynthesis to convert CO_2 from the atmosphere into carbohydrates such as starch and

Top left: Molecular structure of a carbon nanotube containing a caged ion. Carbon can form an immense number of compounds, many vital to life. Carbon occurs free in nature as diamonds, amorphous carbon, and graphite, pictured below.

Carbon circulates in the oceans and in the atmosphere in the form of carbon dioxide. Plants on land and other organisms at sea, such as the algae on these rocks, use photosynthesis to convert carbon dioxide into carbohydrates.

cellulose. These carbon compounds can be used for energy by plants or by animals, energy consumers that eat the plants.

Chlorophyll, the chemical that makes leaves green, is the agent that traps the energy in sunlight and uses it during photosynthesis to make food. In the ocean, photosynthesis is performed by phytoplankton, microscopic plantlike creatures, or algae.

As a by-product of plant photosynthesis, oxygen is released. Animals breathe in the oxygen and breathe out carbon dioxide as a waste product. This carbon dioxide is produced as a result of combustion of sugars in animals' bodies, and is then removed by the blood and released through the lungs.

When plants and animals decompose, the carbon compounds in their trunks, leaves, or bodies are broken down to simpler forms by bacteria and fungi; some of the carbon is stored in soil or fossil fuel deposits or taken up by new plants, and the rest cycles back into the atmosphere.

CARBON SINKS

Carbon that is not circulating in the atmosphere and ocean is stored in Earth's biomass and in rocks, sediment, or fossil fuel deposits. These storage areas are known as carbon sinks. One common carbon sink is calcium carbonate, or limestone. The world's coal and oil deposits are also a carbon sink. Coal and oil are called fossil fuels because

they are made of ancient carbon deposits—the remains of plants and microscopic sea creatures that did not decompose completely after dying in the tropical swamps and oceans of the Carboniferous period, 300 million years ago. Burning these fossil fuels releases CO_2 into the atmosphere that had been out of circulation for all those millions of years.

Cold water absorbs and holds more CO_2 than warm water. As ocean water warms up, it releases some of its stored CO_2 into the atmosphere. CO_2 is also released when forests burn or are cut down.

GREENHOUSE GASES

Carbon dioxide is one of the gases that helps to hold in some of the Sun's radiation and keep the planet's atmosphere hospitable for life in much the same way that a greenhouse traps the sun's heat to keep plants warm when the weather is cold.

Too much carbon dioxide in the air can make the Earth too hot, and thus affect the ability of different kinds of life, including human societies, to survive on different parts of the planet.

Over the past 150 years, carbon dioxide in Earth's atmosphere has risen to its highest level since the dawn of humankind. Part of the rise in CO_2 can be traced to human activities such as burning fossil fuels and cutting down forests. Decreasing the levels of CO_2 in the atmosphere is also within our power.

Solar energy can provide power for homes while cutting carbon dioxide emissions.

HOW TO DECREASE CARBON DIOXIDE EMISSIONS

Use recycled paper products.
Turn off lights and appliances when not in use.
Plant trees.
Switch to compact fluorescent lightbulbs.
Adjust thermostats three degrees warmer in summer, cooler in winter.
Choose cars with high gas mileage.
Choose energy-efficient appliances.
Farm with low-till techniques.
Use renewable energy and green building techniques.

Managing Ecosystems

In a world as crowded as the Earth, with so many life forms sharing the same ecosystems, the planet's inhabitants are bound to get in one another's way. It is a challenge to make room for all the "interest groups" involved in keeping the ecosystem healthy.

Even just among humans, the competition between interests is strong. In the United States, for example, more than a third of the land is publicly owned, and how to manage that land is often the source of bitter disputes and controversy.

For instance, fruit-growing farmers and the tourist industry are both important contributors to Florida's economy. Yet, the water diverted for agriculture and development threatens ecosystems in the Everglades that are vital to the long-term health of all Floridians. Meanwhile, poor building practices, including filling in wetlands or building structures right on the beach, destroy the natural structures that help protect humans from the effects of devastating storms.

Managing ecosystems wisely and fairly is about understanding the various demands on a system. Managers must respond to short-term demands while predicting future consequences and maintaining the long-term sustainability of the system.

Above: Masai with cattle on the Masai Mara National Reserve, Kenya. The Masai are traditionally nomadic and depend on cattle for their economic survival. Their land-management techniques have been used as a model. Right: Florida's Everglades have become a symbol of ecosystems under siege. Top left: A Japanese market. Human population growth is presenting ecological challenges worldwide.

GLOBAL GRASSLANDS

Masai herdsmen in Kenya use land management techniques that keep their grasslands healthy and allow their cattle to coexist on the savannahs of southwest Kenya with a diversity of wildlife. The Masai strategy involves herding on large, unfenced areas owned cooperatively by groups. Texas ranch land, in contrast, has at times turned to desert, as ranchers graze their livestock on limited pasture, leading to overgrazed land and making it difficult to respond to changing conditions such as drought.

Ranchers from Texas and Kenya, united by the similar conditions of their grassland biomes on opposite sides of the world, have come together to learn from each other about both rangeland management and methods of ownership.

Biomes are a large-scale way of thinking about ecosystems. By looking at how one biome evolves, and how it reacts to challenges such as drought,

Preservation of its wild habitat is seen as key to the survival of the endangered Sumatran tiger (Panthera tigris sumatrae).

pollution, human development, and other environmental stresses, we can start to understand much about the ecology of similar regions in other parts of the planet.

SAVING THE TIGERS

When the World Wildlife Fund decided to try to save tigers on the Indonesian island of Sumatra from extinction, they adopted an ecosystem preservation approach, reasoning that a tiger population can't survive in the wild unless it has sufficient healthy wild habitat in which to hunt and hide and mate and to feed its young.

The Sumatran landscape inhabited by the tigers was big enough and sufficiently wild, the organization's

scientists felt, to give the tigers a good chance of reversing their population's decline. Tigers require plants and animals for shelter and prey. Central and southern Sumatra has some of Indonesia's last low-lying rain forest. Unfortunately for tigers, the forest also appeals to loggers and owners of rubber plantations. Another threat is poachers, who sell tiger parts to people who will pay high prices for products made from very rare animals.

The government of Indonesia agreed with the ecosystem preservation approach and created a 212-square-mile (550 sq. km) national park in Sumatra's tiger habitat. Now the job of the WWF and local groups committed to tiger preservation is to make sure people respect the park's boundaries, to enforce antipoaching laws, and to protect resources, such as water, that cross the park's borders but also pass through human settlements.

CHAPTER 5

HABITATS AND LANDSCAPES

Left: Many animal and plant species live in very limited areas. The giant panda is now found only in southwestern China, in mountain forests that support dense stands of bamboo, its favored food. Top: The oceans that surround the Galapagos Islands isolate its populations of animals, such as the Galapagos tortoise, which is now found only on these remote islands. Bottom: Porcupine caribou have adapted to the specific conditions in which they live in Alaska's Arctic National Wildlife Refuge.

Common houseflies are found in households throughout the world. These cosmopolitan travelers adapt to a wide range of food sources, temperatures, and surroundings. Other species, in contrast, live only in very limited areas. Pandas, for example, are endemic to the bamboo forests of western China and are found nowhere else in nature. Similarly, certain pollinators, such as the meganosed fly of southern Africa, have evolved feeding apparatuses that link them to a limited number of flower species.

The physical properties of a habitat have a significant effect on the organisms that live in it. Soil, geography, and climate are key factors in the evolution of native species. The physical limits to how widely a species can disperse may be defined by a geographic barrier, such as a high mountain range or broad ocean, or by competition from other species. Most plants and animals become specifically adapted to the conditions in which they live; these conditions become requirements for survival. Porcupine caribou, for example, in the Arctic need the ground to be frozen at a certain time of year so that they can migrate to winter feeding grounds.

Weather and Climate

Meteorologists have a saying: If you do not like the weather, wait a few days and it will change. But if you do not like the climate, you had better move. Weather is a rainy day, a morning too windless for sailing, a cold front bringing snow. Weather can change daily; it provides sunshine and rain to help plants grow, but also rainstorms that destroy through floods. Climate is the weather averaged over a long period. It is the conditions that can be expected: In summer the climate of Los Angeles will be sunny and dry; in Mumbai the monsoon will come in June, rain will be likely every afternoon, and humidity will be very high.

A RAINY DAY

Rain, snow, sleet, fog, sunshine, clouds, wind, sleet, temperature, humidity—the countless elements that make up weather— all occur in the troposphere, the thin inner layer of Earth's atmosphere that, not coincidentally, also contains much of our biosphere. The geography of a place affects its weather. Weather at the ocean shoreline differs from that on a mountaintop, even if the two places are quite close together.

Plant life also affects weather. A wetland or forest processes water more slowly and stays moister

Top Left: Violent weather can occur along fronts, the leading edges of moving air masses. Above: An afternoon downpour drenches a coconut cart in Mumbai. South Asia is one of several regions on Earth affected by heavy annual monsoon rains.

than a rocky place or a town where much of the land is covered with asphalt. Warm and cold, dry and humid air masses are constantly on the move. This is what makes the weather change from day to day. The leading edges of these masses are called fronts. Fronts are often the location of the more dramatic weather phenomena, such as thunderstorms and gales.

Meteorologists are scientists who study weather. After collecting data from satellites, radars, and sensors such as thermometers (for temperature), barometers (for air pressure), and wind gauges, meteorologists run computer models to help them make their forecasts.

OCEAN AND AIR CURRENTS

Climate is affected by global, regional, and local patterns in the oceans and atmosphere. The coastlines of continents turn

In harsh tundra terrain, flowers, such as this yellow pasqueflower, stay low to the ground.

currents away. Mountain ranges deflect airflows. Sea surface temperatures interact with atmospheric conditions.

Some climate phenomena, such as the interaction of atmosphere and ocean known as El Niño, and the atmospheric circulation pattern called the Arctic oscillation, affect climate in cycles of months or years. Others, such as the jet stream in the atmosphere, and the slow, massive circulation called the ocean conveyor belt, remain constant for millennia. In the latter, warm surface water from the Caribbean tropics, for example, constantly circulates toward the North Atlantic, in the process helping to warm the climate of much of Europe so that it rarely snows in winter in spite of the northerly latitude.

ADAPTATION

Climate influences the ecology of a place by affecting the important physical attributes—water, temperature, sunlight—that determine which organisms can grow and thrive. Over time, the plants and other organisms adapt to their climate. Bulbs in temperate regions hibernate in winter, and then sprout again when the weather warms and the days grow longer. Tundra flowers, growing at high altitudes in harsh conditions, stay short. The life cycles and habits of animals also adapt to the rhythms of climate. A Yellowstone cutthroat trout in Idaho, for instance, has a set of instinctive expectations about April: The snow will be melting, the stream flow high—indicating the right conditions for spawning.

CLIMATE CHANGE AND VARIABILITY

Over long periods, climates fluctuate. Periods of unusual cold, warmth, rainfall, or drought, if repeated for several years in a row, can have a significant effect on food webs. This is climate variability. If such variability persists over a

Predictable weather changes in a climatic zone influence the life cycles of its flora and fauna, such as the breeding cycles of trout that inhabit Idaho's mountain streams.

longer period so that it becomes the norm, it is called climate change. Climate change and variability cause major disruptions for plants and animals that were well adapted to the previous conditions. Organisms that have a hard time adapting to the change may be threatened with extinction if they cannot move. The pressure of a growing human population on Earth's resources also makes it more difficult for plants, animals, and other organisms to adapt to climate change.

The Soil Beneath

Soil is the underappreciated underpinning of life. A soil's chemistry and the organisms it supports are bound up with the unique ecology of a place. In Georgia in the United States, the soil, red from iron oxide, sustains forests of pine trees; England's rich black loam soil, with long, moist growing seasons, produces the country's famous flower gardens. According to Edward O. Wilson, because rain forest soil is so poor, of the 7 million acres of rain forest that have been cleared for agricultural use, fewer than 3 million acres have actually been able to support crops.

The ecology of soil includes both its chemistry and its living components. Chemically, soil can have a high or low acidity (pH) level, which is often controlled by levels of chemicals such as nitrogen and phosphate. Soil organisms live in complex communities, organized into niches, food webs, and specialized roles.

Degraded soil can be improved by adding missing living and chemical elements. These elements may come from decomposed plants in the form of compost or manure, or from petroleum-based fertilizer (which is also a form of decomposed

plants, but fossilized). The plants and animals living on the surface also affect the health of the soil.

WHAT IS SOIL?
Soil is where the important job of recycling dead organic matter into the building blocks of life occurs. Soil is full of life. Mixed in with bits of rock, clay, minerals, and dead plant and animal matter are large numbers of microscopic living creatures— amoebae, protozoa, nematodes, fungi, molds, and bacteria. Many soil dwellers are detritivores, eating dead organic material and helping to break it back

Above: The state of Georgia is well-known for its dense, red, claylike soil, which takes its color primarily from iron oxide. Top left: An English garden at the height of summer bloom. The fertile black loam soil of Britain contributes to the garden's lushness.

An excavated hillside reveals the striated layers that make up an area's soil cover.

A seedling sprouts from cryptobiotic soil, which is common in the high deserts of the Colorado Plateau. These "living soil crusts" are easily destroyed by human activities.

LIVING SOIL CRUSTS

In some places with very little rainfall, such as the high desert in the western United States, the soil surface may develop a delicate living crust made up of cyanobacteria, fungi, lichen, and moss. The U.S. National Park Service calls these cryptobiotic soils "among the first land colonizers of the Earth's early land masses." These living soil crusts not only helped form the planet's early soils, but also helped generate the oxygen that supported the Earth's first oxygen-breathing life forms. When destroyed, often by off-road vehicles or mining exploration, cryptobiotic crusts can take years to regenerate, and meanwhile the soil is exposed to erosion.

down into ingredients needed by other organisms. If you slice a cross section from a hillside, you are likely to see several layers of soil of different colors. Near the top, in addition to bits of rock and clay, will be a concentration of decaying plants, live roots, living and dead insects, larvae, and microscopic organisms. Farther down, the soil will have less organic material and will be more like the rock it came from.

According to the U.S. Bureau of Land Management, on the arid rangelands of the western United States the majority of ecosystem diversity occurs belowground, and up to 90 percent of the total productivity of rangelands takes place in the soil. That means in these areas, far more organisms live belowground than above it.

EROSION

Topsoil, the surface layer of soil in which most plants grow, takes hundreds or even thousands of years to develop naturally from decaying plant material and eroding rock such as limestone. Wind and rain can erode topsoil slowly over long periods of time, or suddenly, as in a landslide after a storm.

In many ecosystems, soil supports enough plant life to control erosion except in unusual circumstances. These ecosystems can be destabilized by severe droughts or overgrazing, however, and topsoil can blow or wash away rapidly, leaving the land unable to support its former plant and animal communities.

It is possible to bring degraded soil back to life, but only with difficulty. Soil ecologists work with farmers to develop tilling, crop management, and other techniques to improve the health of soil, which in turn improves the ecosystems that depend on soil for their health.

Geographic Barriers

Geography can have a profound impact on the evolution of plant and animal species. When mountains, canyons, or currents isolate for long periods two groups of organisms originally of the same species, the DNA of each group evolves differently. When the two populations become so different from each other that if they did encounter each other again they could no longer breed together, a new species has formed.

In western India there are many pressures on elephant populations. The growing human population has pushed the subcontinent's native elephants into several preserves, dotted like necklace beads across what was once their native territory.

Studying the elephants' DNA, conservation biologists made a surprising discovery. Long before humans disturbed their range, Indian elephants as well as other species had divided across a gap passing through a mountain range known as the Western Ghats. These mountains presented such a big geographic barrier that many animals and plants did not cross it for thousands of years. After such a long time of living north of the Palghat Gap or south of it, the elephants on each side became genetically distinct.

Genetic diversity is one of a species' major defenses against difficult times. Shrinking elephant populations can mean dwindling genetic diversity as well as dwindling habitat. The preservation of genetically distinct groups of elephants by the geographic barrier in the Western Ghats helps to maintain a more diverse population of Indian elephants overall. For such a small and endangered population, this diversity could mean the difference between survival and extinction.

ISLAND SPECIES

It is difficult for terrestrial species to migrate onto or off islands, especially islands located far from any mainland. To reach Hawaii, for instance, plant spores had to be blown thousands of miles in the air over open ocean; reptiles had to float on debris. Birds, other than the few species that normally migrate through the islands, had to seriously lose their way.

As a result of their isolation, islands can be ideal places to observe the formation of ecological systems and the way in which new species subdivide from old in response to their isolation and new environment. From

Above: Human population pressures have pushed Indian elephants into small preserves.
Top left: Islands can be ideal places to study the workings of isolated ecological systems.

Madagascar to Tasmania, islands develop species and ecosystems over time that occur nowhere else in the world.

The chain of islands off South America called the Galapagos, which played a vital role in Darwin's development of his theory of evolution, is the only place where iguanas have learned to swim. Finch subspecies on the islands developed a fascinating range of specialized beaks, each suited to its particular diet. By occupying different food niches, Galapagos finches were able to live together in a small space without competing.

Hundreds of hummingbird species evolved to fill specialized ecological niches in the Andes Mountains, such as this one from the Ecuador–Maquipucuna Cloud Forest.

VULNERABILITY

The number of species of plants and animals on an island is directly related to the size of the island. The larger a land mass, the greater the number of plant and animal species it will support.

Unfortunately, the specialized species and ecosystems that evolve in island isolation, adapted to a limited range of predators and competitors, are vulnerable to invaders such as rats and snakes. Native plant species may have a difficult time competing with introduced weeds, which are often naturally selected for hardiness and ability to reproduce easily.

Over the past few centuries, as human populations and globetrotting have increased dramatically, islands, no matter how isolated, have been invaded by foreign species. Geographic barriers that preserved isolation for thousands of years are easily broken down by the arrival of modern ships and airplanes.

ANDEAN HUMMINGBIRDS

Not all species evolve because of physical barriers separating them from their relatives for long spans of time. In some cases, new species evolve because individuals or populations are more successful at reproducing and surviving if they make use of distinct ecological niches, even when they are not isolated by physical barriers.

This is what happened to hummingbirds on the slopes of the Andes Mountains in South America. Some hummingbirds became adapted to rain forest,

Charles Darwin drew the conclusion that Galapagos finches all came from a common ancestor, but had diversified and evolved various beak forms by adapting to local food supplies on the different islands.

others to high mountains or other ecosystems. Today there are more than 300 species of hummingbird, and the greatest diversity of hummingbird species is in the Andes. Researchers have found that for hummingbirds of the Andes, as well as for many other organisms, the driving force in forming new species was the need to adapt to their environments, rather than physical isolation.

Landscape Ecology

Landscapes, like biomes, are defined by common types of vegetation, by physical land forms, and by resources such as water and soil nutrients. Landscapes, more local than biomes, are often studied with an eye toward human activities. Landscape studies incorporate both built and natural features.

One reason scientists study landscapes is to guide land use and resource planning. Since a good part of planning takes place on a local or regional level, the landscape perspective is a useful scale to use. Like ecosystems, landscapes have no clear boundaries. A typical landscape boundary is a watershed, valley, or river delta.

Mapping landscapes, analyzing their patterns and dividing them into their component parts, allows scientists to do things like build models and study how patterns affect resources. With more information about the landscape level, experts can anticipate potential risks and consequences from hazards such as fire, flood, or drought.

If, for instance, more houses are built on a hillside, and woods are cut to make room for housing lots, the hydrology or water cycling through the system will be affected, as well as erosion and bird habitat. Forests filter rainwater. Runoff is slowed as rainwater is taken up by plants and seeps into soil. Paved streets, on the other hand, shunt water rapidly after a rain. How much forest can be cut down before it makes a big difference?

ELEMENTS OF A LANDSCAPE

A coastal area might be considered a landscape, with parts including a wetland, woods, a pond, and homes built into a hillside. Landscape studies have

Above: A golf course along the southern coast of Long Island, New York. Top left: Iguaçu Falls on the border between Brazil and Argentina.

An ecosystem's water cycling is affected whenever trees and vegetation are removed from a landscape and replaced by homes.

been used to understand how pollution will affect life in an area. For instance, they can provide guidance for managing the amount of permissible chemical emissions by factories into streams or farms into groundwater, and the amount of water allowed to be taken out of streams for irrigation. Landscape level effects may be cumulative: The agricultural runoff from any one farm does not by itself affect coastal fish nurseries, but put together, the runoff from cattle operations has a cumulative effect on fish.

Breaking down a landscape into its basic parts helps in the scientific study of landscapes. Scientists like to get a baseline reading on the health of vegetation and wildlife, the nutrient supply and pollutant levels, from which to measure in the future.

Mapping, data collection, and remote sensing are all tools used by landscape ecologists to understand the nature of how a landscape functions.

Scientists provide important guidance about how much water from a farm's irrigation system can be taken out of streams and rivers without harming the surrounding landscape.

SPACE AND TIME

The time period and spatial scale at which landscapes are studied can vary a lot. A landscape can be as small as a backyard or very much larger. This leaves room in the field of landscape ecology for answering both long term, large-scale questions and studying more local, specific issues. For instance, a landscape ecologist could study how beaver trapping in the seventeenth and eighteenth centuries, and the resultant drop in beaver populations, changed streams and water systems in Michigan. A smaller

scale issue might revolve around an industrial landscape in a city. How land use decisions over several decades can be reversed so that the area can again become a place suited to pedestrians.

In Oregon, salmon population is related to both forestry practices—how and where trees are cut—and water resource management. Landscape studies help planners decide how to divide water among farmers, towns, and salmon. Landscape ecology brings together information from many specialties to get a view of a place as a whole.

Wetlands, such as this one in Saskatchewan, Canada, are one element in a landscape.

Oceans and Corals

Oceans cover more than 70 percent of the planet and contain at least as many species as live on land. Marine biologists divide ocean species into three groups: plankton, which are floating, weak swimmers; nekton, which are stronger swimmers such as fish; and benthos, or bottom feeders. Plankton include both photosynthesizers called phytoplankton and microscopic animals called zooplankton. Phytoplankton make up the base of the marine food chain.

Limiting factors for life in the oceans include sunlight, dissolved oxygen, and availability of nutrients. Because of this, most sea creatures live within 30 feet (9 m) of the water's surface. When water is cloudy, sunlight cannot penetrate very far, which means less photosynthesis can take place.

Humans have had a large impact on ocean life. For instance, modern fishing techniques are extremely destructive. Populations of many species of fish have declined precipitously because of overfishing. Some fishing techniques destroy habitat on the bottom of the ocean, where corals, seaweeds, and bottom feeders live.

SEA GRASS

Marine ecosystems along coasts, where seawater comes together with fresh water from rivers and other runoff, are often very productive and full of life. One such ecosystem, the sea grass

Some modern fishing techniques can be extremely destructive to marine ecosystems.

meadow, is an important part of many coastal water zones. Sea grass serves as a nursery for minnows and mollusks, which in turn provide food for others farther up the food chain, including bigger fish, crabs, and birds. Sea grass meadows, with their strong roots help protect coastlines and prevent erosion. They serve to store carbon, and they make coastal marine environments healthier by filtering and cycling nutrients. Sea grass can be damaged by runoff from sewers and fertilizer, as well as by boat propellers.

CORAL AND KELP

Some parts of the ocean have very high biological diversity concentrated in a small area. These places include coral reefs and kelp forests, sometimes called rain forests of the ocean, because they concentrate biodiversity and are also fragile and vulnerable to destruction. Coral reefs are made

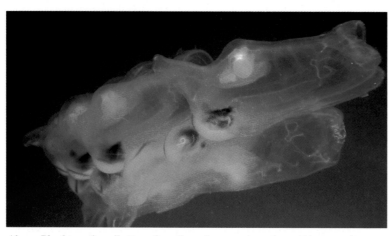

Above: Plankton, the collection of small marine organisms that resides near the water's surface and serves as sustenance for larger marine organisms. Top left: A school of fish. Many once-abundant fish populations have dropped sharply because of overfishing.

by colonies of small soft creatures called polyps, which secrete a calcium carbonate (limestone) house. The brilliant colors of most shallow-water polyps generally come from microscopic algae that live in their tissues. A coral reef has so many niches and crevices that it is a favorite habitat for a great number of fish, anemones, starfish, and crustaceans.

In addition to supporting biodiversity and being a nursery for many ocean fish and other creatures, coral reefs protect about 15 percent of the world's coastlines from erosion. This is particularly important for many island nations that would not exist without their protective coral barriers.

Coral reefs are very fragile because they grow slowly and can be sensitive to changing conditions. The familiar shallow reefs are threatened by warming seas, as well as destructive fishing and tourism practices. Deepwater reefs, which appear to be as brilliant and diverse as their shallow-water cousins yet are only just being explored, are often destroyed by fishing trawlers.

OCEAN REGULATORS

Oceans help to control the climate of Earth in several ways. They absorb sunlight over their large surface area and deliver the heat around the planet on ocean currents. When frozen, as at the poles, they reflect a great deal of the Sun's heat back into space. The oceans receive water that runs off the land and store it until it evaporates back into the atmosphere. The oceans also store a large quantity of carbon dioxide.

A shallow reef off Felidu Atoll in the Maldives Islands. Coral reefs are rich repositories of marine life. Less familiar deepwater reefs are only beginning to be explored.

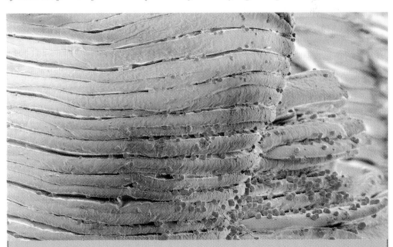

Deep-sea bacteria (blue) on the surface (yellow) of annelid worms (Alvinella sp.).

LIFE IN THE DEPTHS

Recent advances in technology have enabled scientists to send robots to explore the deepest parts of the ocean. In these places, sunlight cannot penetrate and food is scarce.

Surprisingly, creatures have found ways to survive in many places where life would seem impossible. Anaerobic bacteria that live on chemicals boiling up through the Earth's crust have been discovered near deep-sea thermal vents. Some other vent bacteria actually carry out photosynthesis using not sunlight, but the dim glow from hydrothermal vents. These bacteria in turn support a stunning array of worms and other animals. These are the only ecosystems yet discovered on Earth that do not ultimately depend on the Sun's light for an energy source.

Habitat Change

A violent rainstorm passes over a forest. The wind and rain loosen the roots of a tall beech tree near a stream and it crashes to the ground, taking down a few smaller trees and vines with it and opening a narrow clearing.

The day after the rain, sun shines into the clearing and reaches a little sapling that had previously been struggling to survive in the big tree's shadow. Farther downstream, fertile sediment freed by the up-rooted beech is washed by the rain onto a meadow. This kind of everyday small habitat change—a rainstorm that causes death, but provides opportunity for life as well—is an essential part of the natural world. Habitats are constantly changing and shifting in nature. The more permanent, rapid, and large-scale habitat changes that human activity has caused in the past century, in contrast, are threatening many species with extinction.

HABITAT LOSS

According to the United Nations' Millennium Ecosystem Assessment, more than half of the many types of the world's grassland has been turned into farmland, and nearly all the world's ecosystems have been transformed through human actions. Even the seemingly endless oceans have been degraded through pollution, overfishing, destructive trawling, and a warming atmosphere.

Above: A farmland vista in rural Alberta, Canada. Much of the world's grasslands have been converted to agricultural land. Top left: A fallen tree in the forest slowly returns to the soil. The death of one organism may provide the opportunity for another one to flourish.

Coastal habitats untouched by human influence are becoming increasingly rare. Beaches are also building up or eroding in response to waves, winds, storms, and relative sea level rise.

Smoldering forest in Belize. Rain forest loss is a leading cause of species extinction.

What is the connection between human activities and species extinctions? Much of Earth's land is now crisscrossed by roads, built up as towns, suburbs, or cities, or parceled into farms. Forests are altered by logging and clearing for agriculture and building construction. Habitat has been turned into habitat islands. Species that previously were able to move freely to follow prey, escape predators, or track appropriate climatic conditions now find their options limited by the barriers of human civilization.

Not all habitat loss is equal. Wetlands and tropical rain forests may not be able to recover easily because so few of their nutrients are contained in the soil. Destroying coastal wetlands, mangrove forests, and reefs increases flooding and erosion. Deforestation increases levels of greenhouse gases in the atmosphere. As discussed in chapter one, hunting was recently the greatest cause of species extinction, but now habitat loss is an equal or bigger threat.

CHECKERBOARDS

How much habitat is enough to preserve a species? Are numerous small parks as useful as a few large ones? How does fragmentation affect species diversity over time? In order to establish useful goals for preserving habitat for the world's wildlife, scientists need to answer these questions.

Conservation biologist Thomas Lovejoy initiated an intensive research project on this topic in Brazil in the 1970s. Called the Forest Fragments Project, this research has taught many lessons, such as that wind penetrates small forest fragments and dries them out, and that over time the initial effects of cutting a forest into small pieces lead to successive effects. If forest ants are unable to find enough food to survive, for example, the birds depending on those ants will be forced to leave the area.

Conservation groups once thought that merely preserving large areas of wild lands with minimal human presence was the best way to preserve the world's biodiversity. Years of experience showed that conservation plans needed to leave room for people to earn livelihoods and use land sustainably. Plans that did not take human needs into account are difficult to enforce.

CHAPTER 6

STRATEGY AND SELF-SACRIFICE

Left: Some species have evolved in a way that shuns the camouflage defense and favors vibrant colors, such as the yellow-banded poison frog. In these instances, nature has paired brilliant coloring with poisonous animals. It's a sort of warning, signaling predators to look elsewhere for a safer meal. Top: The complex task of finding a food source prompted the honeybee to develop a dance language. Bottom: The web design is hardwired into a spider's brain.

Poison dart frogs don't bother to camouflage themselves but instead stand out in vibrant yellows, blues, and oranges. Bright color has come to mean a warning to would-be predators: Don't eat me. What is it in nature that connects brilliant coloring with deadly poison?

What conditions led spiders to develop a spinneret—a silk-producing organ—and spin orb-shaped webs, the blueprint for which is now genetically programmed into their brains?

Behavioral ecology is the study of how organisms' strategies, choices, and adaptive behaviors are shaped by their fellow creatures and their nonliving environment. Every aspect of an organism's life begs for an explanation. Why do geese fly in formation? How do they choose where to nest? What evolutionary forces led them to pair up for life?

If curiosity killed the cat, as the saying goes, then why is curiosity a successful behavioral strategy for great white sharks? Perhaps it's because they must continually explore new feeding opportunities. Whatever the reasons, an organism's behavior and strategies will be deemed a success or failure in light of its fitness, or ability to produce offspring and pass on its genetic signature.

Foraging Strategy

A lizard basks in the Sun. The longer it basks, the warmer its body grows. The Sun's warmth helps the cold-blooded creature improve its digestion and stimulates its metabolism. But basking also takes time away from the lizard's foraging for crickets and grubs. How much of its time should be spent on basking, and how much on foraging for food?

A mathematical model can be constructed to analyze the costs and benefits of basking versus foraging, to determine the optimal amount of time a lizard should theoretically spend on each. Some other factors might come into play: How much energy is spent on foraging compared with calories taken in? Is foraging more dangerous than basking, taking the lizard to places where it might find another creature foraging—for it?

The lizard, however, will not be able to use a mathematical model. It will instead follow foraging strategies and behavior developed by generations of responses to competition and to environmental conditions—strategies and behavior that have been written into its very genes.

Above: A black scorpion. Top left: An iguana balances the time it needs to warm itself in the Sun with its need to forage for food. Some lizards have developed the unique ability to eat the nutritious, but dangerous and hard-to-find, scorpion.

FOOD AND PHYSIQUE

Ancient ancestors of lizards and snakes waited in ambush for passing prey. Scientists can see from these creatures' jaw and skull structures that these early reptiles stuck out their tongues to catch small prey and must have used visual clues, like movement, to locate it.

Today, lizards employ several foraging strategies. Members of the iguana family tend to sit and wait for prey to come along. Since this foraging strategy does not take much energy, these lizards can afford to have a slow metabolism and take in relatively few calories. They move fairly slowly and may eat prey such as ants and termites that tend to clump in one location.

A second strategy is wide-ranging, active foraging. Geckos move around, actively searching for prey to capture with their rapid, sticky tongues.

Where scorpions live, there are usually also a limited number of lizards that specialize in eating them. Scorpions are nutritious, though relatively difficult to find and eat, requiring special talents in a predator.

Physically, both lizards and snakes have over time evolved hanging jaws that enable them to grasp and bite prey. Many lizards and snakes, especially active foragers, have also developed sensory organs to find their food even when it is hidden and motionless, and to investigate it before eating.

A New Caledonian crested gecko waits to capture prey with its sticky tongue. Geckos have adopted an active foraging technique.

GROUP EFFORT

Bees forage for pollen by scouting individually, looking for flower blossoms. When a scout finds the nectar she is looking for, she returns to the hive and performs a dance that contains detailed information about the distance and location of the food source. The dance even describes how desirable the food source is. Other bees from the hive understand the information by following the directions of the dance. Soon, they fly off to find the source of nectar and bring it back to the hive.

Bees have evolved many cooperative and social behaviors, but only the complex job of foraging for food led these insects to develop a dance language.

Carnivores that hunt in packs cooperate in another way. Wolves and lions and wild dogs, for instance, use group efforts to bring down their prey and feed their young. Lions raise their young cooperatively; all the mothers nurse all the cubs as well as protecting them together in a nursery until they can take care of themselves.

Top: The sleek North American whiptail lizard ranges widely for its food. Bottom: The diet of this desert horned lizard consists entirely of ants. Its stout, camouflaged body and short legs are sufficient to carry out a sit-and-wait strategy to capture its food.

TWO STRATEGIES

These two lizards have developed very different foraging strategies. The North American whiptail lizard is slim and fast, with a bright yellow racing stripe running down its side. The whiptail ranges widely while foraging for beetles, grasshoppers, and termites, and runs to cover when threatened. The desert horned lizard is built like a tank, with a large stomach and short legs. It eats ants exclusively, found through a sit-and-wait strategy. It moves slowly, using camouflage and spines for defense.

Like other carnivores that hunt in packs, lions use group effort to hunt and feed their cubs. Among a pride of lions, the mothers will cooperate in the nursing and protecting of their young.

Passing on the Genes

Individuals that produce the greatest number of successful offspring are the ones whose genes end up in future generations. This means that any genetic trait that enhances an individual's ability to attract mates and produce offspring—looks, behavior, or even chemical cues—will probably get passed on as well, even if it is unrelated to survival or physical strength.

In animals, mating behavior can range from aggression to flirting to, in the case of certain frogs, fertilization entirely in the mother's absence. Each species' ecological setting influences its mate-seeking strategy.

For many plants, seed dispersal is the equivalent of what many marine invertebrates do when they set their young adrift in the water, as if to go find good homes of their own. Some plants—for example, all dandelions in North America—do very well at passing on their DNA without ever engaging in sexual reproduction. Like animals, many plants use appearance, smells, and behavior (though not aggression) to improve their chances of success in passing down their DNA to future generations.

FLOWER REPRODUCTIVE STRATEGIES

Flowering plants coevolved with pollinators, producing, through natural selection, many successful relationships. Flowers will enlist not just insects but birds and even bats to do the job.

Color, shape, timing, location, and fragrance of flowers each act as incentives for some pollinators and deterrents for others. If one flower, whether clover or apple blossom, can attract a bee who visits many flowers of the same species, the chance of pollination increases.

Another strategy to build close relationships with pollinators is the shape-that-fits strategy. The study of biological shapes is known as morphology.

When one flower is a perfect fit with one particular beak or proboscis, both flower and pollinator benefit. The bird or butterfly gets exclusive rights to that flower's nectar, without competition. The pollinator, in return, can feed on only one species of plant.

In South America, close to 8,000 flowering plant species are pollinated exclusively by hummingbirds. Flowers pollinated by hummingbirds are often reddish in color, easy to see in daylight, with a deep tube in the center containing nectar but no fragrance. The deep tube and lack of fragrance both decrease the chance that

Above: Flowering plants have developed close relationships with pollinators, such as birds and bees. A flower's color, shape, and fragrance are among the factors that attract pollinators. Top left: An African honeybee pollinates a flower.

insects will pollinate these flowers, which have evolved to suit the hummingbirds' specific talents.

SEED DISPERSAL

Seed dispersal is another part of the plant kingdom's reproductive strategy. Wind, water, animals, and gravity are common tools plants use to disperse seeds. A dandelion seed parachutes along the breeze. A blue jay eats a berry, then eliminates the berry's seed over a meadow. An acorn falls onto the forest floor, perhaps rolls or is washed downslope, where it remains dormant until spring. Coconuts float away on ocean swells. Some plants also use mechanical dispersal—seedpods that burst explosively when disturbed, for example, flinging the seeds away from the parent.

Some plants have very finely tuned dispersal strategies. The honey locust, for instance, holds its edible seedpods well above clumps of sharp spines. In the environment in which this tree evolved, only mammoths could reach the high seedpods. The honey locust's tough, flat seeds need to be chewed and passed through a herbivore's digestive tract to germinate. Thus they had evolved with the mammoth as their preferred disperser. Yet thousands of years after mammoths went extinct, honey locusts can still be found across North America. They continue to grow intimidating defense spikes, but now they depend on birds, cattle, and small mammals to disperse their seeds. One key to the honey locust's success is its flexibility when it comes to seed dispersal.

Some plants use outright trickery to attract favored pollinators. Flies are attracted to the Rafflesia flower because it emits the scent of rotting meat.

THE ORCHID AND THE WASP

Most flowers provide a nectar reward to their pollinators, but some orchids rely on sheer trickery. The Chiloglottis family of orchids, native to Australia, offers no nectar but sends out a pheromone that mimics a female wasp ready to mate. The wasp rushes to the orchid—which in some cases has even evolved to physically resemble a female wasp—and while attempting to mate gathers pollen, which may then be carried to a different flower, resulting in pollination.

Other plants also use mimicry to attract pollinators. One well-known example is the largest flower in the world, an Indonesian rain forest native called Rafflesia or corpse flower, which emits the scent of rotting meat to attract flies that pollinate it. Mimicry in plants can also be a defense: The lithops, or stone plant, a South African succulent has evolved to resemble a rock in order to survive in a harsh desert habitat where animals eat every leaf or twig they can find.

Some flowers have developed a shape and color that is well-suited to exclusive pollination by hummingbirds. These flowers often are reddish in color, making them easily visible. The deep tube in the flower's center contains a nonfragrant nectar to discourage pollination by insects.

Herd Mentality

A huge school of fish moves gracefully through the ocean. It changes direction in an instant, the fish darting off in a lightning flash. Few of them, maybe five percent, are involved in deciding where the school is going. Most are just trying to stay close to their neighbors.

The wildebeest traveling in vast herds over the Masai Mara grasslands, and the gnats in a cloud around the wildebeests' heads, are both following the same principle of herd mentality.

Keeping close to the group will help protect individuals against predators. The herd members are not helping each other on purpose; every individual is in the herd because it reduces the chance of getting eaten—at the expense of others—and allows each animal to make use of resources found by others, although the entire herd must then share resources.

Scientists have shown that it is not necessary for very many members in a herd to have information about where they are headed or why.

The group gets as much benefit from small numbers of leaders as larger ones. Groups of animals will use as few leaders as possible. The tasks that leaders take on are often more complicated, using energy and exposing them to risk. Followers are less likely to find danger. All they have to do is stick close to others.

Above: Animals that travel in herds, like these wildebeest in Kenya, do so because there is safety in numbers. They reduce the chance of being eaten by predators. Top left: A caterpillar learns little from its parents.

ABILITY TO LEARN

While herd animals can depend on their herd-mates for decisions throughout their life spans, some creatures must start showing independent behavior from birth. From caterpillars to frogs, many creatures never see their parents or have contact with any kind of teacher.

Birds species can have precocial or altricial chicks. Precocial chicks, like those of chickens and ducks, come out of the egg with their eyes open, covered in down, and capable of walking around. Altricial chicks—those of goldfinches and robins, for instance—are born in a helpless state, unable to fly or walk, often without feathers and with their eyes closed.

Each method has its costs and benefits. If one way were "best," all species would do it the same way.

CONDITIONING

Learning is closely related to fitness. In biology, the term "fitness" is used to mean reproductive success—the organism that is able to produce the most offspring is the "fittest." Learning occurs when an organism changes its behavior based on its own experience.

Conditioning happens when a creature knows that a certain

behavior will lead to an expected reward, such as a rat that presses the green button in a lab and expects to receive a pellet of food. In the wild, conditioning is part of life. A rat in the wild will quickly learn where to find the best food sources near its den.

Scientists have found that many animals, even sharks, can be conditioned to act in a predictable manner, and carry a strong memory of the conditioning. They can also learn to associate two objects or states that seem to have no relation—such as when a certain flower blooms, then a certain prey will be available. Successful conditioning to identify food sources and to avoid danger are two very helpful strategies for increasing fitness.

Another kind of learning is imprinting, when a newborn bird or other animal decides that the first creature it sees is the parent it must follow and copy. Goslings usually imprint on geese, but if one happens to see a human being first, it could imprint on that person and ignore other geese.

Cognition, or thought, a third type of learning, requires the brain to be hardwired, or prepared with certain neural networks.

Above: Goslings learn by imprinting. Newborn geese identify as a parent the first creature they see and will follow and copy that animal. Humans possess cognition, one of the three learning types. Certain neural networks hardwired into the brain make cognition possible.

Right: In contrast to imprint learning exhibited by goslings, sharks can learn by conditioning. They associate particular behaviors with a specific reward.

Can Animals Think?

Some male bowerbirds will decorate a place where they hope to attract a female exclusively in shades of blue. Others build tall arches constructed of curved twigs, or create a shiny path surrounding a couple of decorated saplings, a construction known as a maypole bower. The decor of bowerbirds, natives of Australia, is meant to attract a mate, and to make the female feel safe while she checks out her prospects. But is she judging on aesthetics, too? The individual taste expressed by these birds can be astonishing.

"Thinking" in animals is a topic debated among researchers. A heron drops bits of twigs in the water as minnow bait. Chimpanzees sometimes use sticks to pick termites out of logs. Can these animals use information to generate original ideas?

Animals have been documented to have personality types—passive, shy, aggressive: An octopus seems to express jealousy when its visitor pays attention to another fish in the aquarium; one fruit fly may dominate all the other flies on a banana.

Scientists investigate animal thinking partly to understand which aspects of brain function are learned, which are biologically based, and what successful reproductive advantages are created by thought.

MIGRATION

Bird migrations are one of the amazing spectacles of nature. Though migration is dangerous and exhausting, more than 500 of the 670 bird species living in the United States embark on seasonal journeys. In spring they head north to where the food supply

Above: The octopus has been known to express bouts of jealousy. Top left: The male regent bowerbird is known to decorate its courtship area in varying hues of yellow to attract a mate. Can such actions prove that animals can think? If so, what ecological advantages are gained?

is plentiful. In fall they head south to avoid difficult winters. Migrating birds may fly for 90 hours at a time. The arctic tern, champion of migration, travels 20,000 miles (32,186 km) round trip each year.

The physiology of migratory birds is exquisitely sensitive to the seasons. In the weeks leading up to migration, birds begin building up fat stores, sometimes doubling their body weight.

Aside from better food supplies, another ecological advantage of migration is that predators are less likely to become specialists if particular birds are not available year-round to provide a food source.

HEALTH AND BIRDSONG

The 9,700 or so species of birds have an amazing variety of songs and calls. These calls communicate danger, define a territory, or attract a potential mate. The quality of a male bird's song may demonstrate his health and how fit he is to be a father. Sometimes birds seem to sing just for the fun of it. The Australian lyrebird has been shown to have the ability to adapt its song in response to a person playing along with it.

The brain wiring that allows birds to learn songs is genetic, like humans' basic capability to learn language. But the actual song is usually at least partly learned, sometimes with local "dialect" variations and frills, during the first few months after hatching.

There is quite a bit of activity in the early stages of a bird's life—growing and learning to get around and get its own food.

Many bird species will engage in seasonal migrations, at least partly to spend their time where food is plentiful. Another advantage may be that predators are less likely to become specialists.

The chameleon lizard holds the advantage of being able to change color at will to blend in with its environment. Animals who possess this ability take cues from their surroundings.

CHANGING COLOR

Another behavior that has definite ecological advantages is the ability to change color at will. Chameleons use this trick to camouflage themselves to match their surroundings—brown for a brown tree trunk, green when sitting on a green stem.

Some fish change color for camouflage purposes and during periods of high activity. Squid, as well as some angelfish and sunfish, blanch when frightened or excited. Pygmy red octopi turn bright red when disturbed. Changing color is a fine example of cues from environment triggering a bodily, chemical-based response.

Self-sacrifice

In evolutionary terms, it might seem as though creatures that sacrifice their own reproductive lives for the sake of others would quickly disappear from the gene pool. Yet many animals do in fact display self-sacrificing behavior.

In colonies of army ants, thousands of individuals toil their whole lives for the sake of the queen, with little chance of themselves reproducing. Generally, only the queen's genetic material is passed on to the next generation.

Honeybees separately evolved a similar social arrangement and have a similar, self-sacrificing social structure in which most individuals never reproduce themselves, toiling instead on behalf of a related queen.

The origins of such behavior in insects are often linked to the close genetic relationship of the queen and her workers. Because of the way sex is determined in most insects, a female worker shares three quarters of her genes with her sister, but only half with her offspring.

It turns out, though, that relatedness is an insufficient explanation for insect altruism. Other factors, such as the benefits of building a really big nest, must come into play.

REDBACK CANNIBALS

The ultimate story of self-sacrifice is the rather gruesome tale of the redback spider of Australia. A relative of the black widow, the redback is a small spider, with a painful bite, that likes to live in urban or suburban areas. When mating, the male sometimes twists his abdomen onto the female's fangs. In more than 60 percent of spider matings, the male is cannibalized.

Male redbacks succeed in becoming fathers more often when they are cannibalized during mating. Scientists have found that more than 80 percent of male redback spiders fail to find any mate. The self-sacrificial behavior may have developed because there is so little chance of the male spiders' finding a mate more than once, or because they are not physically able to father more than one brood. In either case, the males' extreme behavior shows how one species has adapted to scarce resources—in this case, females—in its environment.

Above: The origins of self-sacrificing behavior in honeybees have been traced to the close genetic relationship of the queen bee and her workers. Top left: A blue jay watches over its territory. Jays display competitive behavior, defending their borders to protect food supplies and mates.

Chimpanzees are the rare species whose actions sometimes appear to be motivated by altruism. When distributing meat after a kill, they will share with other members of the cartload that seemingly had nothing to do with the hunt. But the meat distribution also may be a conditioned response, as the sharing seems to be a way to recognize status and seniority in the group.

SOCIAL INTERACTIONS

Like self-sacrifice, competitive behavior is ultimately connected with contests for scarce resources. If food supplies and mates were unlimited, a blue jay would not have to spend its time proclaiming its territory and defending its borders against intruders of the same species. If there were always a surplus of females, male peacocks would not have evolved their striking plumage and their elaborate tail-shaking mating dances.

Social behavior involves many forms of communication. It ranges from play to fighting, from warning with sound to the elaborate physical displays of mating dances. Most animals, not being very visually oriented, use scent and pheromones to communicate information about potential mates or the location of food. Reliance on pheromones is one reason why many creatures are able to function so well at night.

On rare occasions, animals will act in a way that seems motivated by kindness, such as when chimpanzees distribute meat after a kill, sharing even with individuals who had no connection to the hunt. Chimpanzees' meat distribution appears to be a way of recognizing rank and seniority in the pack. This apparent altruism may instead be a conditioned response to the unpleasant consequences of not sharing.

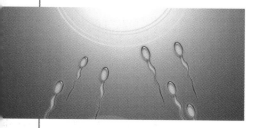

Reproduction Strategies

Organisms have many options when it comes to reproduction. They can clone themselves, either splitting themselves in two or creating a whole new individual from some small portion of their body. They can engage in parthenogenesis, a form of asexual reproduction that can create a little variability among offspring through the process a female uses to create eggs. They can also swap genetic material with another individual, a process we normally refer to as sexual reproduction. Although organisms from bacteria to humans engage in sexual reproduction, the reason for the evolutionary success of this process is still a mystery. One leading theory suggests that sex creates variability that helps populations "outrun" parasites and diseases. Indeed, animals that can reproduce both sexually or asexually are more likely to produce offspring sexually when their environment has lots of parasites.

LIFE HISTORY

"Life history" refers to the path an organism takes in creating the next generation. Even in sexually reproducing organisms, there is a lot of variety in timing of reproduction, the number of offspring, and how much nutrition or care offspring get from their parents.

Above: In early spring young shoots signal a return to life for the Hosta plant. Birth and regeneration of many plants and animals is timed to coincide with better availability of their food supply. Water and sunlight are most available for plants in springtime. Right: A female grizzly bear may mate with several male grizzlies each mating season to maximize the chance of acquiring good genes for her offspring. Top left: An illustration of spermatazoa swimming toward an ovum.

Environment influences the evolution of reproductive strategies. For instance, many animals, as well as plants, reproduce seasonally, so that the birth of their young will coincide with better availability of their food supply (for plants, this would mean water and sunlight).

Environmental change can have big consequences for the relationships between expected

food supply and reproductive timing. Climate change, for instance, can shift food supplies, disrupting reproductive strategies. The pied flycatcher of Britain builds her nest in an oak tree and feeds her babies caterpillars that hatch out on the tree at the same time she hatches her chicks. In recent years, though, warm weather has led the flycatchers' oaks to leaf out early, and the caterpillars to hatch before the flycatchers' eggs do, straining the chicks' habitual primary food source.

The field called life history evolution explores the tradeoffs and correlations that improve an individual's fitness, and therefore its likelihood of having a large number of descendants. There are tradeoffs, for example, between the need for protection from predators and the evolutionary advantages of genetic diversity. Female robins that select a single mate for a whole mating season—or wolves that choose one mate for life—may receive protection and food for their babies from their mate.

Above: A mother baleen whale and her calf swim off the coast of Florida. Whales are known as K-strategists because they have relatively few offspring but invest much time and energy into ensuring their survival. Large animals, such as wolves and humans, tend to follow a K-strategy for reproduction. Besides few offspring, K-strategists are characterized by a long gestational period and the rearing of young in stable environments. The "K" is the mathematical symbol for "carrying capacity."

Above: Black cockroaches are considered r-strategists because they reproduce rapidly. The "r" stands for "rate of reproduction." The high number of offspring for r-strategists helps offset their low survival rates.

A female bear may mate with several males each season, maximizing the chance of getting good genes for her offspring.

Some animals reproduce more slowly, or take longer to reach maturity and be ready to reproduce, in reaction to local environmental conditions. Fish raised in lakes, for instance, have fewer and later offspring when they are crowded.

RESOURCE ALLOCATION

Many scientists consider reproductive strategy to be, in the end, an allocation of resources issue. Biologically, larger babies mean fewer babies; to produce more numerous babies, each one must be smaller and more vulnerable. Yet the probable number of surviving offspring per parent may be roughly similar for the two divergent strategies.

Animals such as cockroaches that have many, relatively helpless babies and spend little energy taking care of them are called r-strategists. Animals such as wolves, whales, and bears, which have few offspring but invest a lot of time and energy into ensuring their survival, are called K-strategists.

The theory of r-strategy versus K-strategy was once thought of as key to defining a species' chances of success. Now, these reproductive strategies are considered to be just two of many resource allocation tradeoffs and issues influencing a species' chances of reproductive success over time.

The availability of good habitat is another important part of the struggle for the success of a species over time. It doesn't matter how many babies are born if those babies do not have a good food source and a safe place to live.

Genetics and Behavior

Charles Darwin postulated that evolution (that is, a change over generations) is inevitable if genetic variation exists within species, and if some variants are more successful at reproducing than others. Like physical traits, behavior has a genetic component. This can be seen from the fact that behavior is often similar for all members of a species.

All robins construct a similar style of nest, but the nest of a robin is very different from that of a weaverbird.

Another piece of evidence indicating that genes play a role in behavior is that similar behavior appears in related species. For instance, fathers are the main caregivers for their young in all seahorse species, but not in any members of the cat family.

Individuals with the more successful behavior will pass more of their genes to future generations, and if the behavior is genetically determined, it will be passed on in those genes. Some small species of east African weaverbirds sleep up to five in a nest. This behavior provides extra warmth for the birds, who live in their nests all year long. Weaverbirds that live communally often build a roof of considerable length stretching across several nests, protecting the group from monsoon rains that could destroy their homes. Within their species, evolution has favored weaverbirds that practice these cooperative behaviors.

HEALTH

Sexual reproduction may have evolved as a way to stay ahead of parasites and diseases. The offspring of peacocks with large, brighter tails grow faster and survive longer, having stronger

Top: Two robin chicks in a nest. Bottom: A weaverbird constructs its nest. The difference between the nest-building styles of robins and weaverbirds indicates that behavior has a genetic component. Top left: A male peacock puts on a colorful show with its plumage in an effort to attract a peacock hen. The offspring of peacocks with large, brighter tails are known to grow faster and survive longer.

immune systems than the offspring of less attractive males. A peacock tail is not just an aesthetic statement but also a signal of genetic fitness.

Female great snipes choose among many males that have gathered together on group display grounds known as a lek. Males give no care after mating, so superior genes would be the benefit of females' choice. In one experiment, male snipes were vaccinated against disease, and females showed a preference for those with higher antibody responses to the vaccination. Without any apparent physical cues, females somehow detected the males with greater immune fitness.

MONOGAMY

Prairie voles are monogamous, sharing nests with one mate for life. Their cousins the montane voles are not monogamous at all. In the 1970s a population biologist named Lowell Getz began a study of prairie voles, now one of the best-documented examples of how genes control behavior.

When two sexually mature and unattached prairie voles breathe each other's pheromones, a genetically programmed chain of events begins. Special receptors in their brains—receptors that do not exist in the brains of montane voles—allow prairie voles to bond with an individual of their species.

Physical changes in prairie voles' brain chemistry correspond to monogamous behavior. Researchers have even found the specific gene that makes prairie voles monogamous and controls behaviors including nesting, defending territory, and caring together for young. Transferring this gene into a montane vole makes this normally antisocial species take on the characteristic mating and parenting behaviors of the prairie vole.

SIBLING RIVALRY

Sometimes, genetically programmed behavior is not so nice. Nazca boobies are gull-like birds that range from Mexico to Peru, with large populations in the Galápagos Islands. Mother Nazca boobies hatch out one or two chicks at a time. If there are two chicks, the first one usually kills the second. Research has shown that the basis for this gruesome behavior is genetic, involving hormones beyond the baby birds' ability to control.

Above: Lifelong monogamy is a unique characteristic of the prairie vole. In the 1970s, biologist Lowell Getz documented how chemical reactions in the brains of prairie vole pairs spur monogamous behavior. Researchers have identified the specific gene that makes these animals monogamous and controls their nesting behaviors. Montane voles, while cousins of the prairie vole, lack this particular gene.

Above: Some animals, such as these Nazca boobies, display rather gruesome genetically programmed behavior. When two Nazca boobie chicks hatch, the first one to hatch usually kills the second.

Ecology Forefathers and Pioneers

Josias Braun-Blanquet
(1884–1980) Influential Swiss-French plant ecologist. Known for his work in classifying vegetation. Developed the Zurich Montpellier School of Phytosociology, an approach to plant community classification.

Rachel Carson
(1907–64) American nature writer and marine biologist. Her 1962 book *Silent Spring* stirred environmental consciousness and led to a ban on the pesticide DDT in the United States.

Charles Darwin
(1809–82) British naturalist whose theory of evolution by natural selection is the central tenet of all biological sciences. Darwin's *The Origin of Species* provided the basis for the theory of evolutionary ecology.

Sylvia Earle
(b. 1935) Marine ecologist and advocate for marine conservation through sanctuaries. Pioneered research on marine ecosystems. Earle's books include *Sea Change*, published in 1965.

Charles Elton
(1900–91) A pioneer of the study of organisms in natural environments, and of animal behavior as part of ecology. Defined the concept of food chains. Published classic books including *Animal Ecology*.

Jane Goodall
(b. 1934) Through more than 30 years of research with chimpanzees in Tanzania, Goodall has been a behavioral ecology pioneer and an ambassador connecting ecological studies with sound public policy.

G. Evelyn Hutchinson
(1903–91) English-born pioneer of freshwater systems ecology, known as limnology. Hutchinson showed how geology, biology, physics, and chemistry interact within freshwater systems.

Alexander von Humboldt
(1769–1859) German student who traveled the world, studying ties between plants and their climate, altitude, and geography. Published *Idea for a Plant Geography* in 1805.

Antoine Lavoisier
(1743–94) French chemist, identified elements including oxygen, hydrogen, and nitrogen. Discovered oxygen's role in plant and animal respiration. Described the law of conservation of matter.

Aldo Leopold
(1887–1948) American wildlife ecologist, author of *A Sand County Almanac,* who argued that natural systems were not only economic resources, but also life-support systems requiring cooperation and care.

Jane Lubechenco
(b. 1947) Trained as a marine ecologist, Lubechenco extended Aldo Leopold's Land Ethic to the sea. Became a leader of scientific advocacy for public policy, particularly on ocean and global warming.

John Muir
(1838–1914) Influential Scottish naturalist and conservationist who founded the Sierra Club and inspired President Theodore Roosevelt to establish the first National Parks in the United States.

Eugene Odum

(1913–2002) Professor who wrote the first ecology textbook, *Fundamentals of Ecology*. Odum's ecosystem approach, focused on the interdependence of all living and non-living elements of a system.

Beatrix Potter

(1866–1943) Best known for children's books, Potter also studied and drew scientific illustrations of fungi and lichen. Her research was presented to the Linnean Society of London in 1897.

Pliny the Elder

(23–79) Roman scholar who wrote a 37-volume encyclopedia on the natural world, called *Historia Naturalis* (Natural History). The books covered topics from botany to medicine to agriculture to geography.

Victor Ernest Shelford

(1877–1968) American community ecologist and first president of the Ecological Society of America. Researched succession in Indiana sand dunes. His studies inspired Charles Elton's work on food webs.

Eduard Suess

(1831–1914) Austrian geologist who invented the concept of a *biosphere* to define the conditions necessary for life in the earth's hydrosphere (water), lithosphere (rock), and atmosphere (air).

Robert T. Paine

(b. 1933) Community ecology; First to suggest the idea of keystone species, using studies on seastars in the Pacific Northwest. Most famous example of this concept may be the otter-urchin-kelp triangle.

Henry David Thoreau

(1817–62) Author of *Walden* and advocate of simple living in contact with nature. Inspired generations of the value of wilderness, and the idea of "living deliberately" in harmony with the land.

Alfred Russell Wallace

(1823–1913) British naturalist and co-creator, with Charles Darwin, of evolutionary biology, and a creator of the field of biogeography. Wrote *The Geographical Distribution of Animals.*

Gilbert White

(1720–93) England's first ecologist. Wrote *The Natural History and Antiquities of Selborne* (1789). Studied nature from field rather than laboratory. Emphasized connectedness of lowly life such as earthworms to all others.

Edward O. Wilson

(b. 1929) Pulitzer Prize–winning ecologist and biologist. Harvard professor. Wrote *The Diversity of Life* among many books, emphasizing importance of biodiversity. A powerful advocate for conservation.

Biodiversity Hot Spots

More than half of all plant species on Earth, and more than 40 percent of vertebrates living on land, are endemic to just 2.3 percent of the land, living nowhere else in nature. When these small areas that support unusually high diversity are lost, hundreds of species, a part of the interconnected fabric that is life on Earth, will also disappear. If these small but ecologically precious biodiversity hot spots can be preserved, then a large number of species may be saved from extinction. The high rate of extinction currently under way is one of the most worrisome issues in ecology today. Some scientists fear that, largely due to habitat loss and climate change, a quarter of all existing species could become extinct in the next 50 years—an extinction rate not seen since the end of the age of dinosaurs. This map shows the location of biodiversity hot spots, as defined by the organization Conservation International.

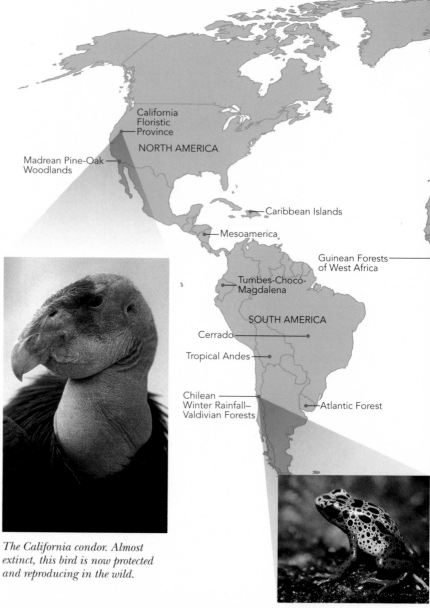

California Floristic Province

NORTH AMERICA

Madrean Pine-Oak Woodlands

Caribbean Islands

Mesoamerica

Guinean Forests of West Africa

Tumbes-Chocó-Magdalena

SOUTH AMERICA

Cerrado

Tropical Andes

Chilean Winter Rainfall–Valdivian Forests

Atlantic Forest

The California condor. Almost extinct, this bird is now protected and reproducing in the wild.

Poison dart frog

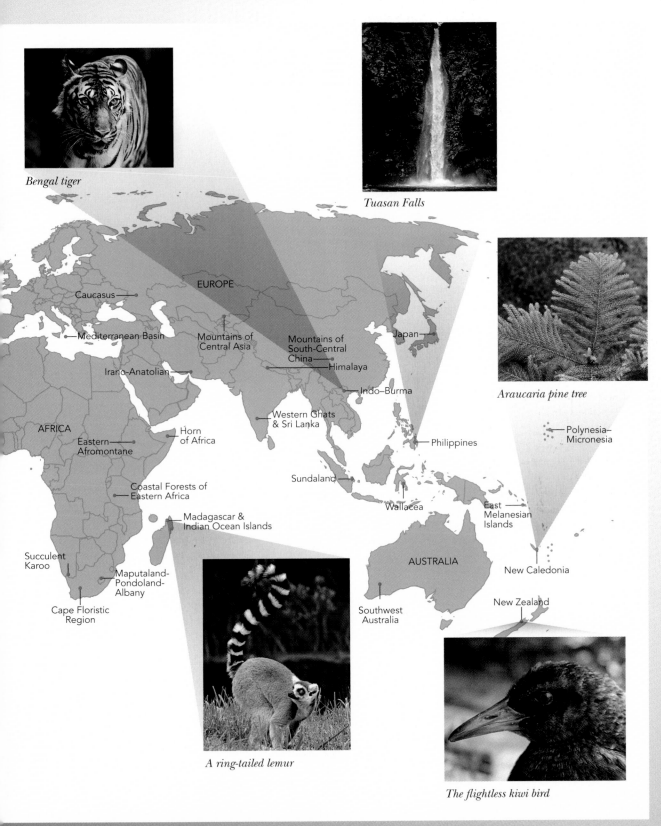

Bengal tiger

Tuasan Falls

Araucaria pine tree

Caucasus

EUROPE

Mediterranean Basin

Mountains of
Central Asia

Mountains of
South-Central
China

Japan

Himalaya

Irano-Anatolian

Indo–Burma

AFRICA

Western Ghats
& Sri Lanka

Eastern
Afromontane

Horn
of Africa

Philippines

Polynesia–
Micronesia

Coastal Forests of
Eastern Africa

Sundaland

Madagascar &
Indian Ocean Islands

Wallacea

East
Melanesian
Islands

Succulent
Karoo

AUSTRALIA

New Caledonia

Maputaland-
Pondoland-
Albany

New Zealand

Cape Floristic
Region

Southwest
Australia

A ring-tailed lemur

The flightless kiwi bird

Climate and Earth's Ecology

Ocean Conveyor Belt

The oceans' thermohaline (meaning heat and salt) circulation pattern influences Europe's warm winters and hurricanes in the Atlantic. Warm water travels north on a surface current, evaporating and growing saltier as it goes. North of Europe, it meets cold water, and the warm salty water, denser than the northern, cools and sinks, mixing with fresh water and returning south on a deep current to complete the loop.

Greenhouse Effect

A blanket of gases, the Earth's atmosphere acts a lot like a greenhouse. Incoming radiation from the Sun has relatively short wave-lengths that can pass through these "greenhouse gases" easily. When this energy hits the Earth's surface, some of it is absorbed by Earth, and later radiated back toward space as heat. This heat energy has much longer wavelengths than the incoming radiation, and the blanket of greenhouse gases prevents it from escaping back into space. As a result, this heat is trapped in the atmosphere, keeping Earth warmer than it would be without the greenhouse gases. This is a normally occurring process without which Earth would be too cold for life as we know it. Human activities have increased CO_2 levels that are likely to raise Earth's temperatures significantly.

THE GLOBAL OCEAN CONVEYOR BELT

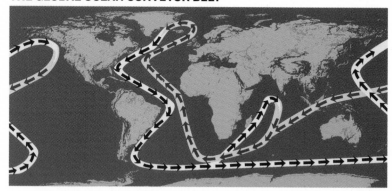

The Global Ocean Conveyor Belt is a circulation system that helps distribute heat throughout the planet. The light-colored arrows represent warm water currents, which travel up from the south toward higher latitudes. Once these warm waters mix with colder northern waters (blue arrows), they sink and move back south to start the loop once again. Any shift in the temperature of these currents causes major climate changes.

1,000 YEARS OF CHANGES IN CARBON EMISSIONS, CO_2 CONCENTRATIONS, AND TEMPERATURE

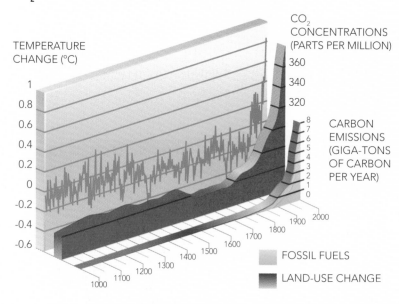

This graph represents 1,000 years of changes in carbon emissions, CO_2 concentrations, and temperature in the Northern Hemisphere, but the findings have global implications. A pronounced increase in carbon dioxide concentrations and a spike in air temperatures, particularly in the last 100 years, make it virtually impossible to overlook how human activities continue to affect Earth's climate.

El Niño/Southern Oscillation

El Niño is a periodic disruption in the normal ocean-atmosphere system in the tropical Pacific. It has global effects and can cause monsoons in Africa and drought in Asia, mild winters in Canada, and drought in western North America. It also affects ocean food webs. First noticed in Peru, El Niño was associated with warm seas, rain, and an absence of fish around Christmastime. (Hence, the phenomenon's Spanish name, which means "boy" or "child" and is a common reference to the infant Jesus.)

El Niño has contributed to significant global climate anomalies, which in turn can lead to flooding such as what happened here along the San Francisco River in Clifton, Arizona. While not all droughts and floods can be traced to the effects of El Niño, which generally lasts 18 months, there are some clear corollaries. The largest floods in Arizona during the twentieth century occurred during El Niño.

Gulf Stream

The Gulf Stream is a warm wind-driven Atlantic current that brings heat and salt from the Gulf of Mexico north along the coast of North America and Newfoundland. Then, influenced by winds, it crosses the Atlantic Ocean to warm Europe. The Gulf Stream's extension, called the North Atlantic Drift, is driven by heat and dense salty water rather than by wind.

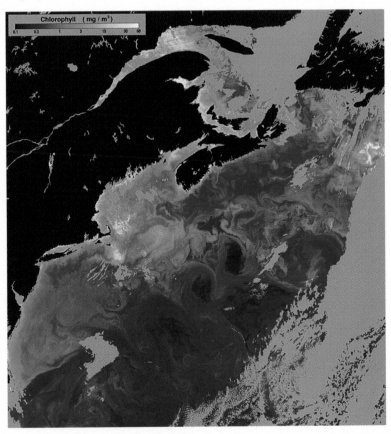

The Gulf Stream is one of the planet's strongest ocean currents and, with its warm waters, a key component of the Ocean Conveyor Belt. This satellite image from April 2005 shows chlorophyll concentration levels. Chlorophyll is an excellent indicator for the presence of marine plant life. The yellow and light-colored areas, where warm and cold currents converge, pinpoint where the concentration is highest.

Arctic Oscillation

The Arctic Oscillation is a cycle involving atmospheric pressure in the northern midlatitude and Arctic. It influences winds, ocean currents, and climate throughout the Northern Hemisphere. The Arctic Oscillation brings warmer, wetter weather to Europe in its "warm phase." In its "cold phase," it brings storms to the Mediterranean and Arctic chill to Europe.

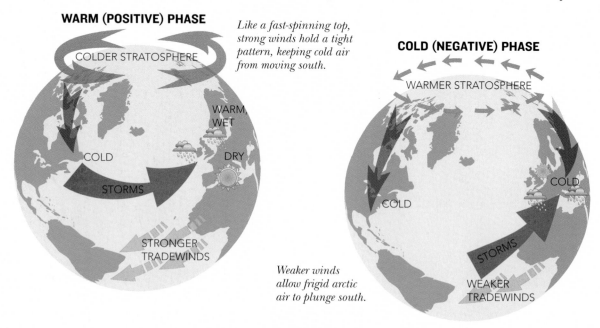

WARM (POSITIVE) PHASE

COLDER STRATOSPHERE

Like a fast-spinning top, strong winds hold a tight pattern, keeping cold air from moving south.

WARM, WET

COLD

DRY

STORMS

STRONGER TRADEWINDS

COLD (NEGATIVE) PHASE

WARMER STRATOSPHERE

COLD

COLD

STORMS

Weaker winds allow frigid arctic air to plunge south.

WEAKER TRADEWINDS

Arctic Oscillation in its positive phase (left) is associated with strengthening of winds circulating counterclockwise around the North Pole (north of 55°N). The fast spinning winds hold the cold air, preventing it from moving south. This causes warmer temperatures and rain in Northern Europe and dry conditions in the Mediterranean and Western United States. The negative phase (right) brings frigid air to Western Europe and the Midwestern United States, and rain to the Mediterranean.

Jet Stream

The jet stream is a narrow air current flowing from west to east at about 20,000 feet in the atmosphere, traveling at about 50 knots, that influences storms and pressure systems at the surface. The jet stream (discovered by World War II bomber pilots) is generated by temperature layers in the atmosphere and the rotating Earth.

A line of cirrus clouds stretching from Sudan west across the Red Sea to Saudi Arabia is the product of a jet stream. Cirrus clouds usually mean fair weather, but jet streams can influence storm systems as well.

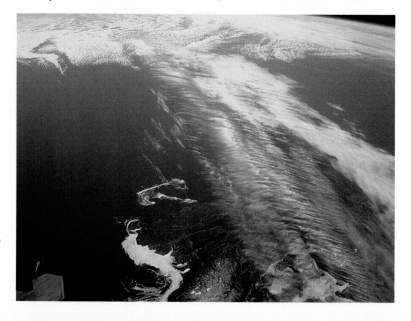

Melting Ice Sheets

Global warming has led to accelerated melting of glaciers and ice sheets around the globe. As ice melts, it forms pools of dark water that absorb more of the Sun's energy than ice, setting up a feedback loop that accelerates the melting. Further, meltwater at the base of glaciers and ice sheets can speed their movement into the sea, also accelerating the melting. One prominent concern for climate scientists is the melting Greenland Ice Sheet, which could dump so much fresh water into the North Atlantic that it would disrupt the Thermohaline Ocean Conveyor Belt. The resulting disruption would wreak havoc with global temperatures.

The Greenland Ice Sheet (right) suffered record melting in 2002. In the middle portion of the sheet pictured here, water pockets have saturated the ice and turned its color from white to gray. Summer melting on the Greenland Ice Sheet generally begins in late April and hits its peak by September.

Arctic Ice and Permafrost

Temperatures in the Arctic Circle are rising faster than anywhere else on the planet, causing the rapid melting of Arctic ice. If conditions remain the same in the future, the ice will continue to melt and permafrost will thaw, causing major changes to ecosystems of the Earth and disrupting the balance of insects, plants, and animals, as well as to the carbon dioxide storage, growth rates, nutrient cycles, and other aspects of Arctic ecology.

A polar bear flees as scientists pay a visit to Alaska for an assessment of the outer continental shelf. As temperatures in the Arctic continue to rise, the entire ecosystem of the area will mutate and affect global climate.

Food Webs

A view of Long Island Sound from the mouth of the Connecticut River along the Connecticut coast. Many valuable wetlands in Long Island Sound have been degraded by contamination and lost to filling, dredging, and "marsh drowning" from sea-level rise and coastal subsidence processes. This turn of events has significantly impacted native wildlife populations and the quality of life for local residents. There have been recent efforts to restore some of the salt marshes along the sound, including the planting of Spartina marsh plants.

The study of ecology is an attempt to understand the interconnections among living things and their nonliving environments. A food web is a diagram of how organisms in an ecosystem depend on one another to get the nutrients and energy they need to live.

The food webs in these illustrations show the three main types of organisms: producers, consumers, and decomposers. Producers are plants, algae, or other organisms that can create energy from light using photosynthesis; another name for them is "autotrophs." Consumers are divided into primary (herbivores), secondary (carnivores), tertiary (top carnivores), and omnivores. Detritivores are organisms such as earthworms, fungi, or bacteria that break down dead material and waste into basic ingredients that can be recycled through the system.

THE FOOD WEB IN LONG ISLAND SOUND

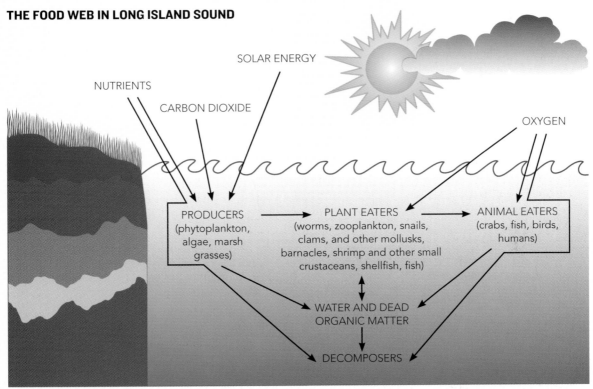

This food web of Long Island Sound shows the order of energy creation and distribution of nutrients in the local marine environment. A food web is a diagram that represents the feeding relationships between organisms within an ecosystem. Food webs generally consist of a series of interconnecting food chains and it is important to understand that they are representative diagrams. Only some of the many possible relationships can be shown in such a diagram, and it is typical to include only one or two carnivores at the highest level.

Detrital food web (salt marsh)

In a detrital food web, most energy and nutrients move from producers (plants, plankton, and algae) directly to detritivores, the organisms that decompose waste and dead material into its basic ingredients. Primary consumers in a detrital food web are creatures such as shellfish and crabs, which feed on the detritivores. Secondary consumers include fish, birds, raccoons, and other creatures that feed on the shellfish and crabs.

Grazing food web (temperate forest)

In a grazing food web, energy produced by plants goes through layers of carnivores and omnivores before eventually falling to the decomposers. When an herbivore eats a plant, much of the energy produced by that plant is lost as heat. It takes a large quantity of producers to support small numbers of carnivores. That is why there are relatively few lions on Earth, compared with grasses and trees.

Marine food web

For reasons that even scientists don't fully understand, the food webs of marine environments tend to be more complex than those of terrestrial ones. One factor may be the greater density of species in aquatic communities. They have more connections among species, and each species tends to eat or be eaten by a greater number of other species. They also have more animals feeding at intermediate levels.

Nutrient Cycles

A carbon molecule becomes part of a tree; when the tree burns, the carbon moves into the atmosphere as carbon dioxide, then gets breathed by a tree again, turned into a leaf, and eaten by a caterpillar. One of the marvelous aspects of life is that all creatures on Earth are formed from a limited selection of nonliving molecules. These building blocks of life include carbon, nitrogen, phosphorous, and sulfur. All these elements move through the biosphere, through living organisms and the nonliving environment (some are stored for long periods outside the cycle, as with oil in underground deposits). Nutrient cycles provide nutrition for living things. Sometimes they are called bio-geochemical cycles, which refers to the fact that they are sometimes part of a living creature, then pass through the Earth, and go through chemical changes as they move through the cycle. Water is another substance that cycles through the ecosystem and is vital to life.

Life on Earth is carbon-based. CO_2 can take many different forms and is found in the air, in the soil, and in the leaves, roots, and trunks of trees and other vegetation. A tree plays a significant role in the carbon cycle, since many of its processes add or remove carbon in the atmosphere. CO_2 is absorbed by a tree during photosynthesis and get dissolved into the soil by sinking, and CO_2 is released into the air by respiration and decay.

CARBON CYCLE

CO_2 IN THE AIR

CO_2 RELEASED BY RESPIRATION

CO_2 TAKEN INTO LEAVES

SOME CO_2 DISSOLVED BY RAIN

DEAD LEAVES FALL

LEAVES ROT, FREEING CARBON

CARBON DISSOLVED IN WATER TAKEN UP BY ROOTS

NUTRIENT CYCLING

A typical nutrient cycle requires three different processes to come full-circle. Through photosynthesis, producers (also called autotrophs) take energy from the sun and convert it to nutrients. Consumers capture these nutrients by eating the producers or other consumers. After consumers and producers are no longer living, decomposers do their part to break down their nutrients to help keep the soil rich in nutrients and sustain producers.

SUNLIGHT

SEA SURFACE

PHOTIC ZONE

ENERGY AND NUTRIENTS

PLANTS

LIVING AND DEAD ORGANIC MATTER

ANIMALS

MIXING OF WATER

DEAD ORGANIC MATTER

APHOTIC ZONE

BACTERIA

DISSOLVED INORGANIC NUTRIENTS

SEDIMENT

Ecological Research Techniques

Top: Divers survey damage to the reef in Fagatele Bay after Hurricane Val in 1992. The reef, located off the island of Tutuila in American Samoa, is home to turtles, whales, giant clams and other sea life. Right: Readying surveillance equipment to detect the nighttime movement of migrating moths.

Field observation

When studying how the world works, there really is no substitute for getting wet and muddy and coming face-to-face with your subject matter.

Sampling

Sometimes understanding can be gained only by collecting and analyzing samples of the pieces of a system, such as water, air, plants, animal dung, or soil. From measuring dissolved oxygen in a river to tracing an oil spill in the ocean, sampling is an important tool for ecologists.

Remote sensing

With remote sensing, equipment automatically collects data and can even send it to people for analysis. Aerial photography was one of the earliest remote sensing techniques. Today weather stations, satellites, and sensors in the ocean collect data that helps monitor the Earth in many ways.

Lab tests

Many ecological questions are best addressed by understanding the chemical and physical properties of a place or creature. Is the water safe to drink? Why are tropical frog populations diminishing? Laboratory tests of field samples contribute to understanding.

Computer modeling

In recent years computer modeling has become an important means of imagining "what if" different scenarios occurred. Computer modeling is especially

Ecologists pull sediment samples out of a lake in Mississippi. These samples are used to study the health of small invertebrates within this ecosystem.

useful for large-scale (either in time or space) questions too difficult or expensive to test through field research.

Bore holes

Drilling cores, sometimes up to two miles long, into Earth's crust or into glaciers and studying the layers of rock and ice provides a snapshot of conditions when the ancient ice or sediment was new.

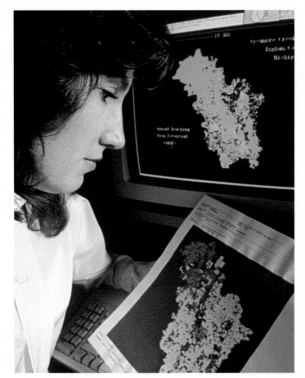

Left: A range scientist uses a computer program to study water samples. Computers have become increasingly important to researchers in tracking and predicting environmental trends. Right: Two technicians stand at the bottom of a three-story tall rainfall simulator while studying how much rainfall the plant foliage intercepts. Bottom: An organic geochemist reaches in to collect a sediment sample that was taken from the ocean floor. These samples help scientists better understand the history and makeup of this hard-to-reach portion of the Earth.

The Ecologist at Work

Lab scientist

Professional ecologists have jobs ranging from college professor to researcher to environmental activist to park ranger. Often, ecologists have jobs in which scientific knowledge and methods of inquiry must be applied to real-life situations. These situations might include helping farmers reduce pollution; balancing competing demands for water; or determining how much development can take place on a site without endangering a nearby natural environment. Ecologists work for engineering firms and government regulatory agencies, as well as for organizations that promote wildlife conservation and stewardship of natural resources. Following are brief descriptions of some of the types of jobs ecologists do. (Many ecologists' work combines several of these elements, such as writing, research, and teaching.)

Natural resources manager
Advises dairy farmers on landscaping to support migratory birds and avoid polluting streams.

Parks employee
Maintains, restores, and monitors health of landscapes and wildlife.

Laboratory scientist
Tests water and mud samples to support research on nutrients and toxins in a coastal ecosystem.

Field scientist
Counts insect populations in a square mile of rain forest. Studies how rain forest insect behavior influences the rain forest and vice versa.

Educator
Teaches about local ecology, introducing the public to wildlife and conservation issues.

Legislative aide
Drafts legislation on public policy that will contribute to a healthy environment.

Computer programmer
Designs models to simulate how crops will grow in various climate scenarios.

Policy analyst
Analyzes difficult issues relating to competing demands on scarce resources and how to provide incentives for wise resource use.

Environmental advocate
Persuades people to make ecologically sound decisions, from saving endangered grizzly bears to switching laundry detergent.

Epidemiologist
Combines medical with ecological knowledge to understand the ecology of disease.

Natural resources manager

Conservation planner

Conservation planner

Defines priorities and locations where conservation efforts should be focused. Tries to balance economic, political, and environmental interests.

Compliance monitor

Helps government authorities ensure that industrial companies are acting within the law to keep air and water clean.

College professor

Researches and lectures on such issues as ice core chemistries or conservation strategies. Trains a new generation of ecologists.

Urban planner

Works with developers and city government to maintain environmental health while encouraging economic goals.

Treaty negotiator

Works with a team to craft careful language expressing his or her country's position on topics such as international fishery or climate change treaties.

Grant maker

Helps a foundation give money and resources to people and organizations working on important conservation and resource management projects.

Writer

Investigates and writes about the most interesting ecological issues of the day. Inspires people to care about the natural world.

College professor

Urban planner

Values and Valuation

Nature provides many services to humans, often free of charge, without which we could not exist. These include pollinating crops, filtering water supplies, watering plants with rain, controlling the climate, and providing wild fish for food. While these services are offered for free, many decisions that affect land use, conservation, and use of natural resources are made for economic reasons. In order to give nature's services greater weight when balanced against the economics of development,

economic valuation of environmental services can be used as a tool for planning. People who oppose placing dollar values on environmental services point out that many environmental decisions are too complex to measure accurately in numbers, and that much of the value in nature is literally priceless.

Dollars and Sense
The examples illustrated on these pages are of situations simple enough to evaluate well in economic terms.

- Preserving the Catskill watershed's natural ability to provide clean drinking water to New York City: $1–$1.5 billion.

- Constructing artificial water treatment plants (not including $300 million in annual operating costs): $6 billion to $8 billion.

- Relocating businesses and infrastructure along the Napa River floodplain in California to prevent flood damage: $155 million.

Top: A bee goes about its business of pollinating a flower. Without this much-needed assist from these diligent insects, many crops would fail. Bottom: Rain falls upon a plant, completing a cycle that helps sustain life on Earth. Plants and trees soak up carbon dioxide in the atmosphere and release oxygen. Right: A fisherman trolls a lake in search of a fresh, healthy, and free dinner. Fish and other wild food sources provided by nature are often taken for granted by humans. For this and countless other reasons, it is of paramount importance to pay close attention to our natural resources and make sure development does not upset the delicate balance that makes up each environment.

A worker performs his duties at a drinking water treatment plant. Construction costs for such a plant dwarf the relatively small amount of money it takes to preserve natural water sources.

- Expected cost over 10 years of repair to businesses and infrastructure after flood damage ($20 million per year): $200 million.

- Amount earned in a single year by a typical ranch, per hectare, operating on clear-cut former rain forest land in Costa Rica: $125.

- Approximate annual fee a rancher could earn from one hectare of rain forest in Costa Rica if paid for the service of removing from 7 to 20 tons of carbon from the atmosphere, as called for by the Kyoto Treaty: $400.

- South African savannah, managed for ranching, income generated per hectare per year: $70.

- South African savannah, managed for ecotourism and hunting, income generated per hectare per year: $300.

Top: A flooded vineyard in Northern California. Heavy rains often cause rivers to overflow, and the damage to businesses and infrastructure can take years to repair. Center: A view of an open-air room at an ecotourism lodge in South Africa. After the abolishment of apartheid, many people won their land back and turned to tourism as a way to make money. Bottom: A commercial beehive facility. Due to dwindling bee populations, some farmers rent hives to pollinate their crops.

- Cost of crop loss in the United States due to declining bee and pollinator populations because of pesticides, disease, and habitat loss: $5.7 to $8.3 billion.

- Annual cost of renting beehives just to pollinate California almond groves: $100 million.

CHAPTER 7

ECOLOGY IN GEOLOGIC TIME

Left: A dragonfly rests on a leaf. While the Earth has undergone vast geologic evolution, the dragonfly is one of a handful of organisms that have remained virtually unchanged. Top: The Margerie Glacier in Glacier Bay, Alaska. Scientists are able to gain information about climate changes by studying elements in rock and ice. Bottom: Sand dunes in a desert, one of Earth's constantly changing features. Many areas that are presently desert may have once been the site of inland seas.

In the vast stretches of geologic time, the entire life span of modern humans is but a moment. If a line were drawn across this page to represent life since the appearance of fish, the first vertebrates, about 500 million years ago, the part representing modern human existence, about 10,000 years, would be only a tiny sliver at the end, less than the width of a human hair.

The science called paleoecology studies the ecology of Earth over geologic time. It investigates not only fossils but also "proxies" in rock and ice, such as oxygen isotopes, that correspond to climate and temperature changes.

It is humbling to think of the changes that a dragonfly, virtually unaltered since the age of dinosaurs, had to adapt to. Oceans have turned to deserts, continents have broken apart, awe-inspiring mountain ranges have been formed, ice ages have come and gone—and the dragonfly has endured.

As human use of Earth's resources tests the planet's capacity to support so much diverse life, researchers have renewed interest in understanding the major challenges that organisms have had to accommodate, and the survival strategies that have succeeded.

Why Dinosaurs Vanished

At the beginning of the age of the dinosaurs, Earth's land surface was all one huge continent, called Pangaea. The land was covered with swampy forests, with relatively few types of plants, among them ferns, conifers, and ginkgoes. Earth's climate was warmer than it is now.

Over the next 100 million years, Pangaea split into two parts. These two segments drifted farther apart and broke up further, until, by the end of the Cretaceous period, continents were more or less located where they are today.

During the approximately 180 million years when dinosaurs

Above: The volcano Mount St. Helens in Washington State. Some scientists believe that the dinosaurs' extinction was caused by a catastrophic event, such as a meteor striking the Earth, or a series of volcanic eruptions. Above left: Ferns were plentiful during the age of dinosaurs.

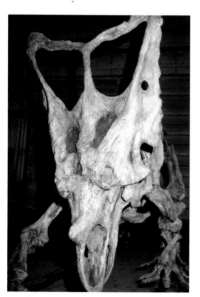

A dinosaur skeleton found in Alberta, Canada. For more than 180 million years, dinosaurs dominated the Earth.

were the dominant vertebrates on Earth, flowering plants evolved, and plant species grew much more diverse. The first modern mammals appeared, as well as many animals still with us today, such as turtles, frogs, and snakes. Many species of dinosaurs evolved and went extinct, while new species appeared to carry on the dinosaur line. Why, after all that time, did dinosaurs—with the exception of birds, which are now believed by paleontologists to have descended from one line of dinosaurs—become extinct?

TWO THEORIES

Scientists have formed two main theories of why dinosaurs, as we think of them, vanished at the end of the Cretaceous period, about 66 million years ago. There is much debate among experts

about which theory is right, but both involve changes in the ecology of the dinosaurs' world.

One theory is that the dinosaurs' extinction was caused by a giant, sudden catastrophe: a huge meteor colliding with Earth, or a series of enormous volcanic eruptions. Vast quantities of soot were thrown into the atmosphere, blocking out sunlight and disrupting the climate. Acid rain and poison gas clouds polluted the atmosphere for years.

A second major theory of dinosaur extinction is that a major ecological shift, part of Earth's natural climate variability, was responsible. Earth was cooling, oceans were receding, and creatures unable to adapt to the changing conditions could no longer survive.

ECOLOGICAL SHIFTS

The two main theories of dinosaur extinction both involve global climate and ecological shifts. Imagine a worldwide shift from a warm climate to a cooler one. Habitats would have been disrupted. If plants died first, herbivores would have lacked sufficient food, and they and their predators would then have starved.

Geological records show that about 60 percent of all species on Earth died along with the dinosaurs. The species that went extinct include the dominant plants on land, but marine life was hit hardest. About 90 percent of algae, the main producers in the ocean food chain, went extinct.

Atmospheric carbon dioxide levels rose rapidly around this time—were there huge fires burning in underground coal fields, or did a die-off of plants cause the rise in carbon dioxide?

While many details remain a mystery, the evidence agrees with what today's ecologists have seen in modern times and come to understand—how changes in a species' ecosystem can lead to its ultimate extinction.

LESSONS FOR THE FUTURE

If the virtual extinction of a group that dominated Earth for 180 million years was caused by climate change and the ecological shifts that accompanied it, the lesson for humans is clear: Practices that could lead to climate change, such as emitting large quantities of carbon dioxide into the atmosphere, are a danger to our continued existence.

A meteor crater in Arizona. This kind of geologic feature yields clues about the cataclysm that may have caused the dinosaurs' extinction.

PALEOECOLOGY

Paleoecology is the study of interactions among ancient plants and animals and their environments. Scientists trying to understand conditions for life on Earth in the distant past analyze rocks and ice, dust, plant spores, pollen, and fossils for clues.

Scientists drill long cores out of glaciers and ocean floors to see what was present in layers deposited millions of years ago. The bubbles of trapped air in ice core samples provide a snapshot of Earth's ancient atmosphere.

Sediment cores taken from the ocean floor indicate that a great change occurred around the time of the dinosaurs' extinctions. Limestone filled with tiny organisms formed the sea floor during the dinosaurs' era, the layers show. But over that limestone layer lies a fossil-free layer of brown clay. Scientists who favor the theory that a meteor caused the mass extinction point out that the high levels of iridium in the clay are unusual on Earth, but common in meteors.

Also, if scientists are able to understand why some rather unlikely species did manage to survive the great extinction— including cold-blooded species such as frogs and snakes—they might learn lessons relevant to preparing humans for an uncertain future.

Many cold-blooded species, such as frogs and snakes, managed to survive extinction 66 million years ago. Scientists can study these species' survival mechanisms to learn valuable lessons relevant to future climate change.

Earth's Early Ecology

The first life on Earth appeared at least 3.5 billion years ago in an environment that had cooled enough both to allow water vapor to condense and form the early oceans and for rock to form a hard crust. Volcanoes had emitted gases to form an atmosphere that included nitrogen, hydrogen, and carbon, but very little oxygen.

Some paleobiologists believe that the leap from inorganic molecules to the first, single-celled, self-replicating bacteria or archaea took place at the site of hot deep-sea vents, because the conditions of today's deep-sea hydrothermal vents seem to be most similar to conditions likely to have existed in the Archean period 3.5 billion years ago.

The earliest life forms were single-celled organisms—archaea and true bacteria. These may have used chemicals such as methane as an energy source, in a manner similar to that of microbes living in the deep ocean and in mantle rock today (survival in today's extreme environments is discussed further in chapter nine).

PHOTOSYNTHESIS

Before the end of the Archean period 2.5 billion years ago, cyanobacteria had evolved that could use sunlight and carbon dioxide in the atmosphere to produce energy through photosynthesis. Oxygen generated as a by-product of photosynthesis at first dissolved in the ocean and combined with iron, leaving behind banded iron formations. Eventually the oxygen did begin to accumulate in the atmosphere, starting the process that would over the next billion years pave the way for oxygen-breathing life.

Oxygen in the atmosphere formed an ozone layer that helped to shield Earth from ultraviolet radiation, making it possible for life to emerge from the ocean and survive.

CONTINENTS AND OCEANS

During the Precambrian period, continents grew larger and formed some of the earliest mountain ranges. As the crust became differentiated from the mantle, tectonic plates formed. Clues to the ecology of early Earth, such

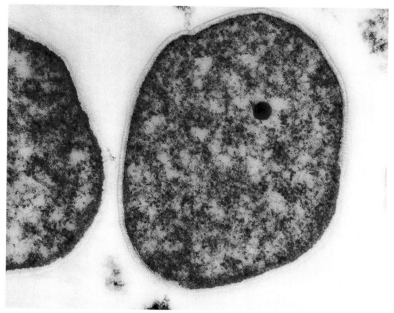

Above: Archaea are believed to be the earliest form of life on Earth, representing the leap made from inorganic molecules to single-celled organisms. Their discovery near hot deep-sea vents has led scientists to theorize that they also were present 3.5 billion years ago when similar extreme conditions may have existed on the Earth's surface. Top left: Life on Earth arose when the Earth's environment had cooled enough to allow water vapor to condense and form the planet's oceans and for a hard surface to form. It is a widely held notion that the Earth's atmosphere was formed out of the gaseous emissions released by volcanic eruptions.

Above: A well-preserved trilobite fossil from the 500-million-year-old Burgess Shale rocks in the Canadian Rockies indicates how far life had evolved by the middle Cambrian era.

forms followed, and as time went on—by about 500 million years ago—almost all the phyla that exist in the modern world had appeared.

A remarkably preserved trilobyte fossil formation in the Canadian Rockies Burgess Shale shows how far life had evolved by the middle Cambrian half a million years ago. The fantastical spikes and frills of these soft-bodied creatures were preserved in detail by an avalanche of fine mud. Fossils in the Burgess Shale include jellyfish and segmented worms, as well as early chordates that were the precursors of all animals with skeletons, including humans. The vast variety of life represented in the Burgess Shale is truly astonishing and demonstrates that the evolutionary paths that led to the life on our planet today do not represent the only possibility.

as the oxygen level of the early atmosphere and the location of landmasses in relation to one another as the plates shifted over time, are detected in records from ice cores and rocks that survive from the planet's early days.

These records show extended periods of mountain building and erosion that formed sediments, which when washed into shallow coastal regions could have created habitats suitable for the proliferation of early marine life. Glaciation during this period extended at times nearly to the equator. Deep-sea hydrothermal vents may have been the primary reservoirs of life.

BURGESS SHALE

The first organisms with a nucleus, called eukaryotes, first appear in the fossil records starting around 2.1 billion years ago. An explosion of life

Top and bottom: The Qaidam Basin on the Tibetan Plateau in Central Asia holds a wealth of clues about Earth's early ecology. Fossils found in the sedimentary layers here date back to the early Miocene epoch (24 million to 16 million years ago) when the plateau was still rising and mountain building was occurring in Western North America and Europe. The planet's tectonic plates were still drifting toward their current positions during this period. For the most part, however, the continental land masses appeared much as they do today.

Mass Extinctions

In the 3.5 billion years or so since life began on Earth, there have been at least five periods during which half or more of the planet's species went extinct. These were at or near the ends of the Ordovician (440 million years ago), Devonian (365 million years ago), Permian (245 million years ago), Triassic (210 million years ago), and Cretaceous (66 million years ago). It is likely that a significant portion of life on Earth also went extinct in the transition from an oxygen-free to an oxygenated atmosphere, although there is no fossil record from this time. Many smaller periods of extinction also took place.

Theories of what might have caused these spasms of loss include major volcanic activity disrupting climate patterns, the configuration of the continents, irregularities in the energy coming from the Sun, impacts from meteors, changes in sea level, and temperature shifts. Not all mass extinctions, however, necessarily had the same cause.

The saltiness of the ocean and the rate at which ocean currents carry warmth from the tropics to the northern and southern latitudes also may have played a role. Another contributing factor could have been methane hydrate, a solid, icy form of natural gas that might have belched large quantities of carbon dioxide into the atmosphere during periods of warming or sea-level drop.

THE GREAT DYING

The Permian-Triassic mass extinction of 245 million years ago was the biggest extinction event in the planet's history. Probably more than 90 percent of all marine species disappeared over a period of one or two million years. More than half of all land-dwelling families went extinct as well, compared to the loss of about 12 percent during the other four major periods of extinction. E. O. Wilson explains that because of the relative rarity of fossils, the unit of taxonomic measure referred to in extinctions is the family, since any one species would be too rare to reliably appear in the fossil record.

Above: E.O. Wilson at left discusses ant species during a nature walk in central Massachusetts. Dr. Wilson, a Harvard University biologist, is among the scientists who have helped to explain evidence pointing to a mass extinction 245 million years ago. Top left: Dinosaurs are believed to have become extinct about 65 million years ago.

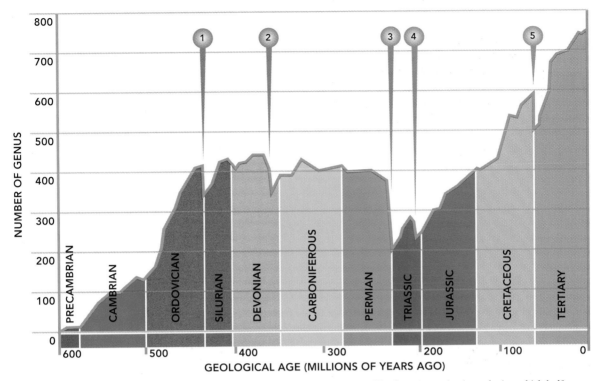

Above: This chart outlines the biodiversity in oceans over the past 600 million years. The five mass extinctions, during which half or more of the planet's species became extinct, are also indicated. Below right: The last mass extinction is thought to have been caused by a meteor impact about 65 million years ago. Evidence of this cataclysmic disaster, known as the K-T Event, was discovered in 1990 off the northern coast of Mexico's Yucatán Peninsula. There, a 112-mile (180-km) diameter ring structure called Chicxulub seemed to fit what would be expected from a meteor impact.

Dominant marine groups that became extinct at the end of the Permian era were the trilobites, the last of the placoderm fish with their bony-plated skin, and eurypterids, sea scorpions up to six feet in length.

Many species of shellfish and arthropods—animals with segmented bodies that include today's insects—also became extinct. One leading theory on the Permian catastrophe is long-term climate change to which species could not adapt.

RATES OF EXTINCTION

The opposite of extinction is speciation, which means the development of new species out of old. Over an extended period of time, it is natural that some species will become extinct while others survive and new species emerge.

During the billion or so years that preceded the Great Dying, there was a nearly 100 percent turnover in species. That is, nearly all species that existed in the beginning of the period became extinct, succeeded by new species.

Fossil studies indicate that the normal, or background, rate of extinction throughout the history of life is between one and ten species per decade

for every million fossil species. Today, given the number of species estimated to exist on Earth, it would be expected that one or two species per year would become extinct.

109

Ice Age Ecology

Ice ages are thought to be caused, at least in part, by small wobbles in the Earth's orbit around the Sun. There have been four long glacial periods in the Earth's history, the first three taking place before dinosaurs appeared on the planet. The latest glacial period started about four million years ago. Over the past two million years there have been about 17 periods when glaciers have advanced and then retreated, or about one every 100,000 years.

The last glacial advance began about 70,000 years ago and ended just over 10,000 years ago. During this period the ice advanced and retreated three times.

In the United States the most recent ice age is commonly called the Wisconsin glaciation, because the mile-thick sheet of ice that extended from the polar ice cap, covering Canada and northern North America, reached down as far as the present-day state of Wisconsin. In Europe the ice reached as far as northern Germany.

Since the ice last retreated, Earth's climate has been in an unusually long interglacial period, with temperatures warmer than they have been for most of the previous 100,000 years. It is possible that we may still be in a glacial period, with Earth's current climate set to retreat into another ice age.

SHAPING HABITAT

Northern North America was dramatically reshaped by advancing and retreating glaciers. Hills were scoured bare, valleys carved. Soil was moved from one place and deposited in another. The course of the Ohio River was changed, and Niagara Falls formed. The Great Lakes were filled with melted glacier water.

Above: Niagara Falls, U.S.A., seen through icicles. Many geologic features, including Niagara Falls and the Great Lakes, were created by advancing and retreating glaciers, which dramatically reshaped northern North America. Top: Earth has experienced four long glacial periods in its history, when much of the planet's terrain resembled this Arctic landscape.

The Baltic Sea at Tallinn, Estonia. Glacial melt at the end of the last ice age mixed with seawater and created a brackish habitat here.

The rocky northeastern U.S. coast from Long Island to Maine, scoured by the ice sheets and redefined by glacial moraine, or earth and stones dropped by the retreating glaciers, differs from the smooth and sandy shore to the south. In another part of the world, the Baltic Sea was filled with a combination of glacial melt and seawater, forming a unique brackish habitat.

Small hills called drumlins, and ridges called eskers, formed by streams on or under glaciers depositing sand and gravel along their beds, are some other land forms created by glaciers.

LIFE ADAPTS

Although the Wisconsin glaciation lasted for tens of thousands of years, living things during that time did not experience an exceptional number of extinctions. Instead, animals and plants seem to have adapted to the changes, no matter how difficult. Insects today, such as dragonflies, closely resemble those of the dinosaur era. Perhaps so few species went extinct because they could easily migrate from one location to another when atmospheric

Mushpot Cave, at Lava Beds National Monument, California. Crystallized plants stored long ago by pack rats allow scientists to study the region's former vegetation.

PACK RAT MIDDENS

Pack rats have a habit of building treasure piles. They stuff plants, bones, and other treasures into caves and rocks in the deserts of the southwestern United States and Central America, then cover them with urine, which crystallizes and preserves the whole package through the ages. Today, by examining pack rat middens, scientists can understand past climates and shifts in plant life. For instance, they can trace how tundra, grassland, and fir and juniper forests retreated at the end of the last ice age and were replaced by desert plants, such as creosote bush, mesquite, and ponderosa pines.

conditions changed. This may also be why birds are the only likely members of the dinosaur lineage that have survived.

Plants, much less mobile than insects, shifted ranges both in latitude and altitude as the climate changed. The plants that lived beside the inland ocean and swamp forests that occupied what is now Utah in the western United States—firs, spruce, and junipers suited to the cooler, wetter conditions that prevailed then—had to move to higher elevations as lowlands turned to desert.

Oceans to Mountains

High in the Himalayan Mountains lie deposits of salt and dark-red fossilized coral, reminders that land now high in these mountains lay under an ocean less than a hundred million years ago, before India migrated north and began shoving into the Asian landmass.

The formation of Earth's mountain ranges, canyons, and other physical features can take tens of thousands, or even millions, of years. Because these geologic events happen so gradually, the vast stretches of time they take is called "geologic time."

The creatures that inhabit the planet today had to adapt and evolve through these constant gradual changes in the Earth's surface during those thousands and millions of years.

SUDDEN SHIFTS

In contrast, major shifts in climate sometimes happen in surprisingly short periods. Five-thousand-year-old paintings in the Sahara Desert show palm trees, hippos, giraffes, and antelope. Animal bones and aquifers under the barren

Saharan dunes provide further evidence that this land was once fertile.

What is perhaps most surprising about the transformation of the Sahara from a land of rich plant and animal life to one dominated by sand dunes supporting very little life is that the change took place within only a few hundred years.

LONG-TERM VARIABILITY

Long-term climate variations on Earth—ice ages are but one example—are thought to be influenced by such factors as small wobbles in Earth's orbit, the shape of its orbit, and the tilt of its axis, as well as cycles of radiation coming from the Sun. These variations occur in cycles of 10,000 years or even hundreds of thousands of years.

An ice core drilled at Lake Vostok in Antarctica reveals that for most of the last 250,000 years, Earth has been considerably colder than it is today. In fact, the relatively warm weather experienced on Earth for the last 10,000 years looks like an unusual interlude when compared with the 200,000 years that came before.

Ten thousand years is a short period in geologic terms, but it encompasses just about the entire span of modern human civilization.

Above: Part of the Himalayan Mountains in Nepal, where geologists have discovered deposits of salt and fossilized coral. This land mass once lay under water. Top left: The Sahara Desert was transformed from fertile land into desert in only a few hundred years.

VARIABILITY AND SURVIVAL

Short-term variations in climate are thought to be caused by climate-driving phenomena such as the El Niño–Southern Oscillation, or ENSO, which shifts ocean currents every few years in ways that cause drought and flooding in many parts of the world. Climate scientists say that dealing with short-term variability is good practice for facing the more difficult challenges of long-term climate change.

Climate variability over periods of even just a few decades can have devastating effects—the Mayan culture may have been wiped out as a result of an extended drought—but species as a whole generally do not go extinct because of these difficult periods.

The chances of a species' survival are affected by many variables, such as genetic diversity, which decreases the likelihood that one disease or challenge will wipe out every individual at once. Diversity also supports evolutionary adaptation. An ermine turns white in winter and brown in summer to be camouflaged from predators— an example of adaptive genetics in action.

Throughout the millions of years since life first evolved on Earth, the constant, gradual, and sometimes not so gradual climate fluctuation has reinforced the need for flexibility.

A flooding, icy river (top) contrasts sharply with a dry lake bed (bottom), whose cracks reveal the effects of drought. While some climate change is short term and variable, longer-term or sudden climate change can have devastating effects. Genetic diversity, and with it the ability to adapt, is one way that species are safeguarded from extinction.

Succession: A Long Story

The Hawaiian Islands as we now know them—for this volcanic chain has been constantly changing over vast stretches of geologic time—began to form millions of years ago as barren humps of hardened basalt arising from beneath the waves. Each island was then colonized by plants and animals drifting through the air and water from the mainland or other islands, becoming over time a rich ecosystem, teeming with life.

The process was gradual: Wind and rain weathered rock over thousands of years, forming sand that would become the basis for soil. Algae and lichen (itself a close symbiosis of algae and bacteria) grew on the weathering rock, providing the first organic material to enrich the soil. This eventually created a suitable habitat for microorganisms and small insects.

Over the next thousands and millions of years, plants, insects, birds, and mammals floated, flew, or swam their way to the islands. Once there, they evolved into thousands of new and unique species and subspecies.

In small local climate pockets, different groups of species thrived and became dominant. Hawaii's many, varied habitats developed, where a snow-covered mountain could tower over rain forest and palm-studded beach.

The first organisms to colonize a new habitat are called pioneers. These are the organisms that help modify the environment to make it suitable for other life. Thousands of years after their birth as heaps of basalt, the Hawaiian Islands possess a thriving, diverse ecosystem.

SECONDARY SUCCESSION

Nature is constantly undergoing and recovering from disturbances both small and large. While primary succession, the

Left: Lush vegetation on the Hawaiian Islands forms part of a rich ecosystem, teeming with plant and animal life. Right: Black basalt geologic formations, arising from beneath the waves. The Hawaiian Islands were originally formed from such barren humps of rock; the process of life taking hold here was gradual. Top left: Aerial view of a bog, where water and land are in constant flux.

development of life in a lifeless landscape, may take thousands or millions of years, secondary succession takes place all the time, over the course of decades, years, or even months. Secondary succession defines the recovery of life after a natural or man-made disturbance to an ecosystem.

An example of secondary succession occurs when agricultural land is abandoned: Over several years it will proceed through predictable stages of milkweed and bramble, to a "second-growth forest" of dense but narrow-trunked trees.

Secondary succession occurs after a variety of disturbances: after a flood washes away trees and deposits silt, or a fire burns through a swath of forest. In each case the land will recover relatively quickly and in stages. Small, hardy weeds and grasses will spring up first out of the mud or charcoal, to be joined by flowering perennials, then sun-loving shrubs, some of which may eventually be shaded out by trees.

CLIMAX ECOSYSTEMS

As organisms coexist, the soil, plant, fungi, herbivore, and carnivore populations continually adjust. Eventually, ecosystems may reach a state in which the abundance and dispersal of life are relatively stable.

These relatively stable ecosystems—though life within them continues to grow and die, evolve and adjust—are called climax ecosystems. Climax ecosystems have a relatively large biodiversity and as much productivity as the soil and weather can support.

An example of a climax ecosystem is the boreal forest of Canada and Siberia, a mix of fir and spruce that, if left undisturbed by man, would probably remain a similar forest until the next ice age, or until global warming so changes the temperature and rainfall patterns that new species can dominate the old.

Arctic tundra is another example. Because of its temperature extremes, tundra cannot support most plant life. That plant life will remain a low-to-the-ground mix of grass, moss, small and hardy flowering plants, and lichen, coexisting with animal life adapted to the harsh conditions.

The ancient ruins of a town in Kenya are being recovered by the forest, in an example of secondary succession.

The coastline of Lake Michigan. Ecologist Henry Cowles has suggested that the variety of plant life here is due to a withdrawing ice sheet, which left behind some Arctic fauna.

INDIANA SAND DUNES

At the end of the last ice age in North America, glaciers left till, or jumbled rocks, around the banks of Lake Michigan. Water and wind acted on the rock, wearing it into sand.

In the early twentieth century, ecologist Henry Cowles noticed something odd about the sand dunes near Lake Michigan. Plants belonging to several ecosystems grew close together: Prickly pear cactus could be found close to Arctic bearberry, southern dogwoods near Northern jack pine. Cowles reconstructed the plant succession, figuring that some Arctic plants remained as the ice sheet withdrew, and plants that had been pushed south reappeared. In some places only tough prairie grasses could survive, but oak woods grew up where deeper soil remained. Wetland flora thrived where an iceberg had been left behind to melt slowly. Cowles found the area to be a virtual museum of succession dynamics.

The Hand of Man

About 50,000 years ago, before the waning of the last major glaciation of the Pleistocene, humans walked across the Bering land bridge from Asia and colonized North America from Alaska to Patagonia within a few hundred years.

These modern humans possessed tools, strategy, and fire. They began almost immediately to act as a keystone species, affecting ecosystems both by killing animals and by reshaping the physical environment.

THE OVERKILL HYPOTHESIS

Many scientists believe that one of the early results of the spread of *Homo sapiens* was the extinction of many large land animals.

North America was home to saber-toothed cats, mastodons, ground sloths, mammoths, long-horned bison, and giant beaver. Fossil records, however—marked both by bones and a notable increase of charcoal bits from fires—show that within a few thousand years of man's arrival, nearly all the largest mammals on the continent had become extinct.

The same thing happened in Australia, where flightless birds, giant kangaroos, and dozens of other species became extinct shortly after the appearance of man.

There are a few theories of why humans drove so many animals to extinction. Perhaps the large animals, never having

met humans before, were unafraid and had developed no defenses. Another theory is that humans hunted so many different kinds of animals that they were not overly affected by the disappearance of any one species.

Some scientists believe that the huge herds of buffalo that roamed the plains when Europeans first arrived were an "outbreak population" that developed when humans eliminated species that would have competed for the plains habitat.

AGRICULTURE

As the last Pleistocene glaciation receded, 14,000 to 12,000 years ago, glacier melt and warmer weather would have allowed many plants, including ones edible to humans, to thrive, to the detriment of cold-adapted plants and animals. Life would have become easier for prehistoric humans. Archaeologists have found evidence that humans at this time, though they still lived as hunter-gatherers, formed settlements around places where animals would come together.

Within a few thousand years, not long in geologic time, humans in the Middle East began to cultivate wild grains. Some researchers think that the need to cultivate grains was spurred

Above: A North American bison, one of the continent's oldest remaining species. Fossils show the existence of many other large land animals, which disappeared a few thousand years after man's arrival. Top left: Fire, man's powerful tool, contributed to the extinction of many species.

by the abrupt mini–ice age called the Younger-Dryas event, 12,000 years ago. Humans who had become accustomed to easy living on wild grains, the theory goes, suddenly found that their food was not growing as well in the colder conditions.

Cultivation allowed for selection of those grains that could grow well in cold. That is, farmers could keep some of the fattest grain heads to use as seed for the next year. The cultivated grains and resulting stabilized food supply allowed humans to settle in one place and start to build cities. Early domesticated plants included melons and dates, almonds and lentils, as well as barley and wheat.

Sheep and goats were domesticated by about 10,000 years ago. Man's goal in domesticating animals was to produce more of the kinds that would serve human needs. Sheep, goats, and later cows provided clothing and food. Oxen and horses helped with the work. The populations of these chosen animals were artificially increased by humans.

The population of predators such as wolves and coyotes, which were seen as competing with humans for food, was controlled through hunting.

The story of man's influence on Earth's ecology had entered a new stage.

Right: Yoked oxen working on a farm. Domestic animals were initially bred for food, clothing, and work; the gradual increase in their population has further altered Earth's ecology.

A wheat field in northern Portugal. The cultivation of grains, some scientists believe, became a necessity when the climate changed abruptly 12,000 years ago, in a mini–ice age. Humans could no longer rely on plentiful wild grains, and learned to grow their own.

WHAT DO ECOLOGISTS DO?

Left: Rachel Carson (1907–65), a zoologist and marine biologist, is widely credited with having launched the modern environmental movement. In the wake of Silent Spring, *her landmark book on pesticides and their harmful effects on humans, scientists and the American public rallied to call on the federal government to impose restrictions on the use of dangerous chemicals. Top: Containers of herbicide wait to be sprayed on a farm field. Bottom: Agricultural pesticides led to the near extinction of the American bald eagle.*

According to the Ecological Society of America, what all ecologists share, whether they are experts in ant behavior or resource management, is a passion for observing, analyzing, and asking questions about the natural world.

In 1962, Rachel Carson, a marine biologist with a long career at the U.S. Fish and Wildlife Service, changed hearts and minds across the nation when she published *Silent Spring,* one of the most famous environmental books ever written. Carson aimed to show that humans are part of the natural world, as vulnerable to damage as other elements of the ecosystem. "The more clearly we can focus our attention on the wonders and realities of the universe about us, the less taste we shall have for destruction," Carson wrote.

Silent Spring's message raised awareness of how the environment could be damaged by widespread pesticide use. Eventually, Carson's warning led the United States to ban use of DDT, an agricultural pesticide that threatened birds of prey, including bald eagles, with extinction.

While few environmental scientists are able to make as big a difference as Rachel Carson did, many enjoy being part of a larger mission: to improve humanity's relationship with the planet we inhabit.

Understanding Acid Rain

In the early 1960s an ecologist named Gene Likens and his colleagues were studying stream ecosystems in the White Mountains of New Hampshire when they were surprised to discover rain with an unusually high level of acidity.

Typically, the pH level of rain would be around 5.2, but the rain that Dr. Likens collected had a pH level of 4.05 to 4.1. The team knew this represented a threat to the environment: Stream dwellers, including frogs, salamanders, trout, and crayfish, can die when pH levels fall below 4.5. Plants, including trees and food crops, also suffer when their environment becomes too acidic. The phenomenon of acid rain had been named a century

before in Manchester, England, and understanding had been advanced by scientists in Sweden, but little work had been done in North America.

The scientists in New Hampshire investigated what could be causing high acid levels in rain falling on a New Hampshire forest far from towns or industry, and eventually traced the problem to upwind burning of fossil fuels, including coal and gasoline.

Specifically, the acid rain was caused by emissions from cars, factories, and electric utilities. Coal-fired electric plants were creating smoke that rose into the atmosphere in the Midwest and fell as acid rain hundreds of miles away in New England.

SCIENCE AND POLICY

For years after the ecologists' research traced the source of acid rain to fossil fuel emissions, state and national governments, environmental protection groups, and industry groups argued over how to solve the problem.

Technology was invented to scrub air as it leaves coal-fired electric plants, and to cut sulfur and soot emissions from car engines. Some lawmakers and environmental groups wanted to force electric utilities to cut their emissions by passing laws limiting the amount of sulfur allowed into the atmosphere. Some utilities opposed new laws and technologies, because they would make energy more expensive or less profitable.

Above left: Amphibians may be more at risk from the effects of acid rain. Above right: Crayfish rely on a hard outer shell for protection. A high level of acidity interferes with their ability to grow this shell. Top left: The White Mountains in New Hampshire. In the northeastern United States, the rain tends to be highly acidic and leads to a reduction in the diversity of organisms.

Above: Researchers believe that two elements, sulfur and nitrogen, are primarily responsible for the harmful effects of acid rain. These elements have a number of sources, both natural and human-made. Contributions from natural sources, such as volcanoes and lightning, are dwarfed by those from human-made sources, including power plants, cars, and other gas-powered machines.

The U.S. government first passed the Clean Air Act in 1963. It was amended substantially in 1970 and again in 1990, when the government set limits on allowable levels of sulfur dioxide emissions.

Polluters could trade allowed sulfur emissions, so that if one company needed to emit more than its allotted limit, it could pay another whose emissions were lower. Since 1990 sulfur dioxide emissions from power plants have decreased by about one-third, according to the Environmental Protection Agency.

The "cap and trade" system was instituted only for sulfur, not some other sources of acid rain, such as nitrogen oxide and ammonia. Deposits from these sources have continued to rise. Europe, whose 1979 Convention on Long-Range Transboundary Pollution similarly controlled sulfur, did later address nitrogen oxides and ammonia as well.

RESTORATION

For decades, ecologists have studied the effects of acid rain. After understanding the cause, researchers worked to quantify effects of acid deposits on organisms ranging from sugar maple trees to lake dwellers. Research also focused on understanding how acid is processed through water and soil.

The chemistry of systems affected by acid deposits turned out to be rather complicated. For instance, alkaline (the opposite of acid) calcium carbonates in the soil can neutralize acid deposits. As it neutralizes acid, though, the amount of calcium decreases, so this benefit eventually ends.

An additional problem caused by acid occurs when it raises the level of aluminum in lakes. Fish cannot tolerate aluminum, and suffocate. These fish deaths are a result of acid deposits, even though aluminum is what actually kills them.

Recently, ecologists have embarked on efforts to study how ecosystems damaged by acid deposits can be restored. Understanding rates of natural restoration, and the costs of various measures that can be taken to help undo previous damage, can affect policy toward electric utilities, industry, farms, and car drivers who continue to pollute.

Ecologists examine the relationship between the environment and actions that affect it. While all ecologists spend time doing fieldwork, some of it rather tedious, research is only a small part of the daily life of these scientists.

FIELD SURVEYS

At some point in their careers, usually starting when they are still students, most ecologists spend time in the field, engaged in sometimes tedious tasks such as counting snails, collecting water samples, or analyzing soil chemistry. Laboratory work suits some, while others move on to look for a job influencing policy, or helping a township or a company make decisions that are ecologically sound. Even well-known environmental scientists whose public lectures and books influence nations continue to conduct basic research on organisms or ecosystems. Marine biologist Jane Lubchenco, for instance, travels the world speaking to people about the importance of caring for ocean life, but she also runs a laboratory at Oregon State University where she researches plant-herbivore interactions, biogeography, and the ecological effects of climate change.

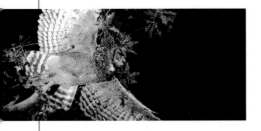

Saving Endangered Species

About 1,000 species in the United States are listed as "endangered." These species are considered in immediate danger of extinction. Several thousand more species or populations are listed as "threatened" or "species of concern" because their populations are declining or their habitat is being lost or damaged. Since the Endangered Species Act was passed in 1973, only seven species in the United States have been taken off the list because they are no longer endangered.

Worldwide, scientists have identified certain locations with unusually high concentrations of endangered species under threat of extinction. If these "hot spots" can be protected, then a large percentage of species will be protected as well.

Many international organizations, such as Conservation International and the World Wildlife Fund for Nature, are concentrating their efforts on these special places.

LAWS AND COMPLIANCE

Ecologists play an important role in understanding wildlife populations. Data gathered about these populations help in determining whether and why species are endangered. Scientific data also allow planners to determine what can be done to help a threatened population recover.

When studying an endangered species or subspecies—whether it is a trout, harpy eagle, or golden lion tamarind—environmental scientists assess population size, range, diet, behavior, and habitat needs, and what is causing that population to be threatened.

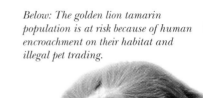

Below: The golden lion tamarin population is at risk because of human encroachment on their habitat and illegal pet trading.

Above, left to right: Humans are the main predator for many animals, including rhinoceroses that are ambushed at watering holes, whales that have been hunted for centuries, and the Harpy eagle. Top left: Spotted owl. Many of the world's endangered species are victims of pollution, poaching, and habitat destruction.

Laws, regulations, and tax incentives are some tools that can be used to protect species. In some cases, however, even a law or regulation is not enough protection. Cutting down rain forests may be illegal, but ranchers in Ecuador may do it anyway because they think the profit to them is worth the risk. Some environmental scientists work in the area of compliance, monitoring conditions to make sure that rules and laws are being followed.

In the United States many federal agencies, including the Fish and Wildlife Service, the National Parks Service, and the Geological Survey, as well as state and city agencies, employ ecologists.

HABITAT PROTECTION

Often endangered species require a large area of wilderness for habitat. This may put these creatures in conflict with humans who have other ideas about how to use land and natural resources.

In 1990, the northern spotted owl was listed as threatened under the Endangered Species Act. The owl's habitat, unfortunately, was in timber country, specifically old growth forest in the Pacific Northwest, 80 percent of which had already been logged.

Logging operations in spotted owl country were both the basis for the economy of many towns and the greatest threat to the last remaining 2,500 pairs of northern spotted owls. With pressure from both conservation organizations and those who live by logging, lawmakers had to decide how much forest to keep

Heavy logging operations, which lead to deforestation, have threatened the survival of many species like the northern spotted owl in the Pacific Northwest.

off limits to timber companies in order to keep the spotted owl from going extinct. Still, this species remains endangered.

Battles over preserving endangered and threatened species often are fought by industry or economic groups that will be hurt in the short term or that believe their long-term viability will be jeopardized. Environmental groups must serve as the "voice" of wildlife that cannot argue on its own behalf.

As an alternative, some groups are working to develop ways for people to earn incomes while preserving endangered ecosystems. Developing markets for sustainably harvested forest products such as wood or nuts, or paying higher prices to organic coffee growers, helps people make enough money without destroying ecosystems. Ecotourism has also become a major economic factor in helping to preserve wild places, although not all ecotourism is harmless.

UNINTENDED CONSEQUENCES

To protect sea turtle eggs from raccoon predators, scientists often place a galvanized metal mesh cage over the eggs after they are laid. But researchers recently determined that the metal in the cages changes the magnetic field around the nest. Because the turtle's internal navigational system is based on Earth's magnetic poles, the change in magnetic field could affect the turtles in unknown ways. This is an example of how even simple interventions can have unintended consequences.

Complex policy decisions must often deal with a bewildering array of possible consequences. Helping one party can often harm another. Ecologists, whether they work in government or industry, must balance scarce money, time, and competing interests. An ecologist working as a water resource manager, for instance, must weigh the interests of everyone who uses local water: farms, factories, homeowners, golf courses, hydroelectric plants, as well as wildlife. When drought reduces water levels, how much should be released from a reservoir? Balancing interests is difficult because someone is always left unhappy.

Regulations and Compliance

Environmental regulations affect some industries more than others. The oil industry, paper companies, waste management firms, mining companies, electric utilities, and manufacturers tend to be heavily regulated. These industries by their nature have an environmental cost. Companies in these industries often hire ecologists and environmental scientists to help them comply with regulations, or in some cases to take a proactive approach to environmental responsibility.

Often, environmental compliance information must be made available to the public. Utility companies are required to file emissions reports, called "toxic release inventories." Some companies are more forthcoming; an Ohio utility also posts on its Web site discussions of its forestry and wildlife restoration programs, to demonstrate that it is a responsible environmental citizen.

Environmental scientists are needed by industry, government, and watchdog agencies to insure that rules and laws are being heeded.

IMPACT STATEMENTS

Before construction, mining, manufacturing, road building, or other development projects are allowed to proceed, developers are frequently required to file environmental impact statements. These are legal documents that provide specified information about how a project will affect the environment.

Information required on environmental impact statements includes how much energy, water, space, and other resources the proposed development will require, and how a project complies with local zoning and land use regulations. If the project will have negative effects on the environment, the developer must suggest ways of making sure those effects are minimized.

Freeway construction in Arizona. Due to the significant environmental impact, U.S. environmental regulations require companies to minimize the negative effects of their projects.

Above: An offshore oil field. Top left: A fuel refinery's smoke stacks crowd the skyline.

Will a project fit into a community physically and aesthetically? Will it bring more people and increased traffic to an area with narrow roads or few schools? Will it be an eyesore? Will it release sediments or pollutants into a river? Open a forest to logging? Federal, state, and local governments all require different information be included in an environmental impact statement.

Environmental impact statements are often produced by engineering firms, or by specialized consultants. Environmental scientists may be hired by these firms to gather data and prepare the reports, by government agencies whose job is to evaluate and monitor new development, or by conservation groups working to limit environmental disturbance.

VOLUNTARY LEADERSHIP

Some companies make sensitivity toward the environment part of corporate culture and image. Ben & Jerry's ice cream, for instance, sometimes decorates its stores with stories of how the company has helped the environment. The company not only donates money to environmental organizations, but also attempts to address the effects of its manufacturing and business on the environment by recycling more than half the waste it produces each year.

Computer chip manufacturer Hewlett-Packard decided to make its newest U.S. plant environmentally friendly in order to save money. The Texas factory uses far less water and electricity than older factories. Both the building and manufacturing process were designed for low environmental impact and high energy efficiency.

Environmental specialists in fields from architecture to printing are contributing to a new consciousness in industry. Lumber companies in Canada collaborated with environmental groups on a plan for managing the last old-growth temperate rain forest in North America, the Great Bear Rainforest, the only habitat for the rare, white "spirit bear," a subspecies of the American black bear.

Where industry and environmentalists used to sit on opposite sides of many arguments, they now find plenty of opportunities for collaboration.

Top: Paper products at a recycling center. Bottom: The Great Bear Rainforest in British Columbia, Canada, is the last old-growth temperate rain forest in North America. Lumber companies and environmental advocates have collaborated on a plan to manage this forest.

Disease Ecology

Traditionally, public health officials and medical doctors have approached disease from the perspective of preventing its transmission and curing those who are sick. Another approach is to look at the ecology of organisms that cause, host, and spread a disease.

Disease organisms are part of nature like every other living thing. Recently, ecologists have contributed to research on diseases from AIDS to West Nile virus, sudden oak death, bird flu, and Lyme disease. Disease ecologists research how disease organisms live and how they fit into their environment.

Lyme disease, for instance, depends on interactions of ticks, deer, white-footed mice (which carry the bacterium that causes Lyme disease), and other small animals. The disease spread to humans as more homes were built in the woods of the northeastern United States, reducing populations of noninfected small animals such as chipmunks and lizards, and increasing contact between ticks and humans. Tick babies (nymphs) increasingly relied on infected mice, then carried the bacterium to humans via deer, whose population exploded as both predators and human deer hunting declined.

Ecologists have contributed to the effort to stop the spread of viruses. Their approach centers on the environment of disease-causing organisms. Top: A blood sample is drawn from a house sparrow. Above left: A deer tick. Above right: A deer with her fawn. Lyme disease's spread partly depends on the interaction between ticks and deer. Top left: A mosquito is a carrier of vector-borne diseases.

Many viruses, such as the lethal AIDS and Ebola viruses, exist for long periods in wild animal populations without causing outbreaks in humans, then suddenly jump from their wild host into humans. One theory that weighs why viruses jumped from monkeys living in rain forests in central Africa to humans is that human populations encroached on the organisms' natural forest environment. This might be a cautionary tale for the future, a warning that humans could unleash more deadly diseases if they keep expanding into the world's rain forests.

ECOLOGY AND PREVENTION

Especially virulent strains of flu occasionally jump from birds or pigs to humans, killing dozens of people and raising worries that bird flu could spark a global pandemic of deadly flu in humans.

Global health organizations and poultry farmers have reacted by killing large numbers of poultry in flocks where cases of flu show up. Health authorities have also kept watch on wild birds. Some dead birds have tested positive for West Nile virus and others for avian flu. As a result, some authorities have recommended killing wild birds, or cutting down trees where wild birds roost, to discourage spread of bird flu or West Nile virus between domestic and wild bird populations, or between wild birds and humans.

Opponents of this plan point out that wild birds are also important members of their ecosystems, and that killing them, unlike cull-

Top: Millions of infected chickens have been killed in an effort to stop the spread of avian flu. Bottom: Virulent strains of flu could make a jump from pigs to humans.

ing domestic bird flocks, could lead to bad effects in natural environments. Some also point out that poor treatment of birds in industrial poultry farms can cause stress to birds and lower their immune resistance to virus.

UNDERSTANDING CAUSES

Disease ecologists often study vector-borne diseases, which are carried by a "middleman" such as rats, mosquitoes, or ticks. Organisms that carry a disease to others are called "vectors."

In the case of West Nile virus the host is birds, and the vector is a certain species of mosquito. In Lyme disease, the host is small mammals and the vector is black-legged ticks.

Vector-borne diseases can be fought by controlling the infected host species or the vector itself. Thus, when health officials are concerned about West Nile virus, they may mount a special mosquito-spraying program.

Widespread programs to kill mosquitoes or ticks are opposed by some on ecological grounds. Even the loss of mosquitoes can be harmful: They are an important food source for many animals. Back in Rachel Carson's day, pesticides used against mosquitoes were so deadly, and used in such a broad manner, that they also caused the death of birds and pollinators such as butterflies and bees, disrupting agriculture.

Field Ecology

Field surveys and experiments are a critical component of ecology. Ecologists study changes in individuals, populations, communities, and ecosystems through both space and time to reveal how organisms interact with one another and their environment. Researchers may watch what happens naturally or, more commonly, manipulate various aspects of a community or ecosystem to better understand how all the parts function. For example, the U.S. Bureau of Land Management recently set up a field survey in Wyoming to assess the impact of two kinds of disturbance—grazing animals (domestic and native) and managed burns—on vegetation communities, especially sagebrush, and wildlife use of habitats.

To perform this study, the government used several classic field ecology techniques. The first is a transect. In this technique an observer follows a set path and records findings, such as sage grouse droppings, according to their frequency along the path. The second technique is a point count, in which an observer counts some variable at a set location—observing, say, how many songbirds show up on a hill grazed by 50 cattle compared to one grazed by 25 cattle. The third technique is live trapping, in which traps are set out at intervals, and the number of small mammals caught in them is an indication of the population of small mammals in the area.

A fourth classic ecology survey technique is mark-recapture. In this technique, a number of individuals are marked at one place and time, and then researchers return to the same place at a different time. For mobile animals, researchers can estimate the actual population size by noting how many individuals they capture on their second visit are marked. For nonmobile animals and plants, researchers can measure how long the individuals live, how

healthy they are, and other variables. Mark-recapture can also be used to measure movement patterns, if the marker provides information both on the original location of the animal and a way for anyone who recaptures the animal to report its new position. Salmon populations may be studied with mark-and-capture techniques, for example, to find

Above: A salt marsh in Queen Anne County, Maryland, is ready for planting. The poles indicate transects used for vegetation surveys. Transects are one of four classic field ecology techniques scientists employ to study ecosystems. Top: A firefighter monitors a prescribed burn in Utah's high desert. Top left: Managed burns have unmanaged effects on vegetation communities.

out how water quality and dam management affect the number returning each year.

Surveys allow ecologists to document patterns in nature. To understand the processes underlying the patterns they observe, ecologists perform experiments, manipulating aspects of a system. An often-used ecological concept, that of the keystone predator, arose from a series of experiments in which Robert Paine of the University of Washington compared community composition of stretches of shoreline where he had removed a predatory sea star with stretches of shoreline with normal levels of sea stars.

A tag is removed from a Chinook Salmon at the Leavenworth National Fish Hatchery in Washington State. Researchers track the fish and use this data to assess the health of native salmon stocks. This type of study, known as mark-recapture, is a classic ecology survey technique.

MESOCOSMS

Sometimes ecologists use small re-creations of ecosystems, with real organisms in them, as a way to conduct research on communities and ecosystems in a controlled manner. These "worlds" created in the field or laboratory are called mesocosms if they are intermediate in size (such as a greenhouse or open-topped enclosure), or microcosms if they are small (like a bowl or a small aquarium).

Microcosms and mesocosms enable complex ecosystems, particularly systems made up of interacting microorganisms, to be studied in the laboratory. Very complex research questions such as how biodiversity affects ecosystem functions can be studied in large scale or by using mesocosms and microcosms.

Researchers might use a mesocosm, in this case looking like a large aquarium, as a way to test the effect of a pollutant (such as a pesticide or industrial waste) on a coastal estuary. While standard laboratory tests would look at the effect on individual species, mesocosms allow researchers to look at the effect on a community of organisms. Rather than asking whether the pollutant is safe for snails, bacteria, fish, and so on, mesocosms address the questions of how that pollutant affects all of them together. Even if a pollutant is harmful to snails, for instance, snail populations may increase in the presence of the pollutant if its effect on the snails' predator is stronger than its effect on the snails.

Setting up several identical mesocosms and adding differing, controlled amounts of the pollutant to be studied allows researchers to conduct experiments that simulate real-world situations such as toxic runoff

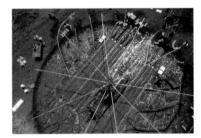

A rainfall simulator in South Dakota is an example of a mesocosm, allowing ecologists to assess water runoff and soil samples.

after rains. Mesocosms can also be used to study, for example, the effects of diversity on ecosystem function, since the diversity of organisms in each tank can be carefully controlled. Soils and soil microorganisms are often the subject of microcosm-based research. Though easily studied in a laboratory, soil is somewhat difficult to observe in its natural setting, as it is not entirely visible to an observer without disturbing the land.

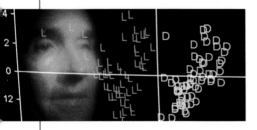

Modeling
and Molecules

Though useful for ecologists, field studies are labor intensive. Experiments that involve laboratory research or remote sensing technology provide different kinds of information.

Mathematical models allow scientists to conduct research that is difficult to do in the field, either because an ecosystem is too large, with complex variables, or because the researcher wants to test several ideas quickly without the time and expense of a field study.

Models are essentially a series of equations that describe how the pieces of an ecological system relate to each other. Models allow researchers to imagine different scenarios—drought, fire, failed food supply for herbivores, for instance—and test how the system would react.

For example, researchers studying the Asian longhorn beetle invasion of northeastern U.S. forests devised a model to explore the life cycle and method of dispersal of the beetle and how populations of maple, elm, and poplar trees have been affected by the infestation. The model helped scientists design a strategy to limit the beetle invasion by surrounding outbreaks and destroying individual infested trees before the larvae in them reach maturity.

MOLECULAR TECHNIQUES

When populations split off and become separate species, or gain characteristics that help them to succeed, molecular ecology allows scientists to read when and where these developments happened.

Before DNA analysis techniques were invented, fossils and observations about distribution and characteristics of living species were ecologists' main sources of information about the history of living things. DNA "fingerprinting" now allows ecologists to trace organisms' family trees with increased accuracy, revealing great and small evolutionary steps.

DNA studies also can inform some present-day behavioral research, revealing, for example, that female sand lizards who mate with many different males in one season have healthier babies, and that female pheasants who appear to be in a single male's harem are actually unfaithful to him much of the time.

Recently, scientists developed techniques to extract and amplify the DNA from animals' dung. Sloughed-off skin cells from the digestive tract, they found, were a reliable source of high-quality DNA. DNA from elephant dung, for example, revealed that a known population of Indian elephants actually was composed

Above: A scientist records sounds produced by Asian longhorned beetle larvae as they feed within an infested willow tree. Researchers studying the beetle infestation have devised models to gather data in order to develop strategies to counter the beetle's spread. Top left: A scientist analyzes texture data for rice samples as part of a mathematical model.

of two distinct species, separated by the Brahmaputra River. Such information is important for conservation efforts.

Like mathematical models, the DNA-from-dung technique helps ecologists simplify field work, in this case eliminating the necessity of capturing, sedating, and taking a blood sample from an elephant before being able to study its DNA. This is especially helpful in studying rare, elusive, or aggressive animals.

REMOTE SENSING

Remote sensing instruments include satellite, sonar, and seismographs. They are used in the same way as, say, a baby monitor that allows parents, while in a different room, to hear when their infant wakes from her nap. Research subjects involving remote sensing can range from land-use patterns to forest cover to storm damage. After large-scale natural disasters such as the December 2005 Indian Ocean tsunami, satellite images are often used to assess ecological damage.

Satellite-based images make it much easier for scientists to study large-scale ecology such as vegetation productivity and climate change, or to do physical mapping. The images can also be processed to reveal some degree of detail about vegetation.

Right: Researchers at the Long-Term Ecological Research Network site in New Mexico's Jornada Basin reset weather equipment used to study global climate change, desert ecology, and the threat of desertification. Top right: DNA samples recently revealed there are two species of Indian elephants.

International Treaties

Environmental challenges can cross national boundaries and affect Earth's common places, such as oceans and the atmosphere, that belong to no one and everyone. International issues such as ocean fisheries regulation and climate change cannot be regulated using laws belonging to just one country.

As a result, many of the world's biggest environmental threats are now addressed with international treaties. International treaties are complicated. Each one is negotiated by hundreds of people from countries all over the world, each bringing different economic and political interests to the table.

The science on which international environmental treaties are based is often evaluated by global organizations such as the United Nations.

PROTECTING THE OZONE

The Montreal Protocol on Substances that Deplete the Ozone Layer enacted in 1987 caused a major change in the way refrigerators were manufactured and made aerosol spray cans relatively rare.

A layer of ozone in the planet's upper atmosphere protects Earth from much harmful ultraviolet radiation. In the 1980s, scientists discovered that the ozone layer was thinning, largely because of man-made chlorofluorocarbons (CFCs) and a few other industrial substances.

The World Meteorological Organization, which monitors the ozone layer, warned of dangers of a thinned ozone layer: "The incidence of skin cancers, cataracts, and infectious diseases among humans would increase; agricultural yields of certain crops would fall; many manufacturing materials would weaken prematurely; and ecosystems would destabilize."

In just a few years, scientific evidence and political will built up, and more than 100 governments agreed to eliminate the use of chemicals harmful to the ozone layer, except in cases where they could not be replaced.

FISHERIES

Many of Earth's major food fish stocks are depleted because of overfishing, habitat destruction, and bycatch (those fish or other organisms that are caught unintentionally). The greatest decline is in large fish such as bluefin tuna and swordfish.

With improvements in technology and increasing human population, many kinds of fish are being caught faster than they can reproduce. Industrial fishing operations often use techniques

Above: UN headquarters in New York City. The United Nations formed the Conference on the Human Environment in 1972 to encourage partnership between nations in caring for the environment. Top left: In 1974, aerosol sprays were found to include CFCs, which do damage to the ozone layer.

Clockwise from top left: Overfishing is a major problem for Canada; human activities such as manufacturing result in the greenhouse gases that have been found to contribute to climate change; U.S. Vice President Al Gore and Japanese Prime Minister Ryutaro Hashimoto at the Kyoto Global Climate Conference in 1997; a plant succumbs to prolonged drought.

CLIMATE CHANGE

Like other issues addressed by international treaties, climate change crosses borders, and no individual nation can control it.

Climate scientists have determined that the world is growing dangerously warmer, and that a portion of this climate change can be attributed to human actions. Fossil fuel emissions are the biggest problem, and the one that humans could do the most about.

In 1997, representatives of more than 160 countries met in Kyoto, Japan, and hammered out the start of an international agreement limiting emissions of greenhouse gases that contribute to global warming. Although the United States and Australia eventually decided not to ratify the protocol, enough nations did that the Kyoto Protocol has become an international force.

(multilateral) have been signed. Unfortunately, enforcement is difficult, and in many cases each fishing boat has a strong economic incentive to take out as many fish as possible, even if it means breaking the rules.

Some scientists believe that the best way of protecting ocean fish stocks is to establish large parks, or marine reserves, where no fishing would be allowed, enabling ecosystems and fish populations with a chance to recover.

that destroy ecosystems on the ocean floor. Because much of the ocean lies in international territory, international treaties are a major means of regulating fishing. Hundreds of treaties between two countries (bilateral) or between several countries

CHAPTER 9

SURVIVAL AT EARTH'S EXTREMES

Left: Plants utilize both physical and behavioral adaptations to survive in the desert. Some are able to store and conserve water or to grow long roots, which allows them to tap into water sources located a distance away. Others remain dormant during the driest periods, springing to life when water becomes available. Top: Penguins are well-suited for the icy barren stretches of Antarctica. Bottom: The Antarctic codfish is found in abundance in the freezing seas that surround Antarctica. The fish is able to withstand the cold because its blood is loaded with glycoproteins that depress the freezing point to below that of the surrounding seawater.

Life exists even in Earth's driest deserts and deepest oceans. Scientists are still working to understand the mechanisms by which plants and animals survive severe dehydration, extreme cold, high levels of salt, and other environmental challenges. In recent years researchers have uncovered life where scientists previously thought none could exist. Among the most astonishing discoveries were lively ecosystems clustered near hydrothermal vents in the deep ocean, where water pressure is crushing and no sunlight penetrates. Sunlight and oxygen used to be thought of as requirements for life, but no longer.

Photosynthesis does occur, using light produced by organisms, but most microorganisms produce energy through a different process called chemosynthesis. It seems that the better our instruments of study, the more life we find, even deep within Earth's rock mantle. Some organisms that survive in hostile environments may be similar to the organisms that lived on Earth billions of years ago, when the planet's atmosphere had no oxygen and a chemistry very different from that of today. Life in Earth's extreme environments increases our understanding of the basic conditions required for life and of how life might someday be found to exist on other planets.

Atacama Desert and Alpine Climates

The Atacama Desert in northern Chile is the driest desert in the world. It stretches 600 miles (966 km) south from the Peruvian border. Near the coast, fog makes up for low levels of rainfall. In some places plants, such as cacti and ferns, have adapted to receiving all their moisture from fog. Recently, humans have begun to build fog-catching nets to collect moisture from the air.

No rain has fallen on the center of the Atacama since record keeping began 150 years ago. The central Atacama is one of Earth's few absolute deserts—a place where no life at all can survive: no cacti, no insects, no lizards, no algae. The central Atacama looks like nothing so much as the surface of Mars. The same reddish sand predominates, the same small rocks lie strewn on the ground, and in every direction the sterile landscape is apparently devoid of life.

At the edges of the absolute desert, where the land still looks as lifeless as Mars, the threshold is crossed, a bit of moisture gathers, and life can again exist. Under the stones, where the little moisture from the air can condense and linger in the shade, life has a toehold. A greenish caste on the underside of rocks here indicates the presence of cyanobacteria (also called blue-green algae, although they are not algae), those tough pioneers.

In deserts around the world—both the cold desert of Antarctica and hot deserts such as the Atacama—algae and bacteria survive under translucent rocks, often if not always quartz, where condensation traps a little water, and enough light gets through to allow photosynthesis to occur. Scientists studying the borders beyond which life cannot exist, and scientists trying to determine whether life might exist on Mars, have looked to the Atacama for clues.

Above: The Valley of Death at the center of the Atacama Desert is the driest region on Earth. It may also be the most lifeless. Inset: Tiny lichen on an Atacama rock is highlighted in false color to show the concentration of chlorophyll. Top left: Animal bones lie on the Atacama Desert floor.

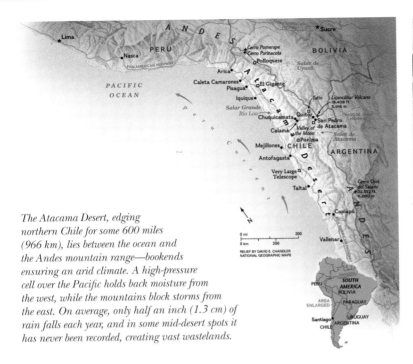

The Atacama Desert, edging northern Chile for some 600 miles (966 km), lies between the ocean and the Andes mountain range—bookends ensuring an arid climate. A high-pressure cell over the Pacific holds back moisture from the west, while the mountains block storms from the east. On average, only half an inch (1.3 cm) of rain falls each year, and in some mid-desert spots it has never been recorded, creating vast wastelands.

When the climate grows warmer, species such as the North American mountain goat will move to higher ground.

ALPINE HIGH LIFE

Because climate changes rapidly with elevation, plants found on mountaintops are quite different from those down the slopes. As elevation increases, forested slopes give way in the space of only a few hundred yards to alpine tundra, with its short, tough vegetation, and mats of lichen on rocks. The species on top are sometimes smaller, tougher versions of lowland species but often are completely different from the life at lower elevations.

Alpine environments generally have short growing seasons, poor soil, little rainfall, high winds, low oxygen levels, and relatively weak protection from harmful ultraviolet rays. Life at high altitudes must adapt to conditions of relative scarcity and hardship. Fortunately for organisms that do successfully adapt, the competition in these difficult conditions is relatively scarce. That is one reason the last twisted pines surviving at high altitudes, whether in Japan or Utah, can grow to be hundreds of years old.

When competition does arise, though, mountain species are backed into a corner. While living things in many environments migrate to escape periodic difficulties such as drought or predators, mountain animals and plants often have no place to go if conditions get worse.

Animals and birds living at high elevations must adapt to relatively low oxygen in the air, a scarcity of water, as well as differences in metabolism.

At elevations where forests abruptly give way to tundra, treeline ecosystems represent the edge of the envelope for tree survival.

CLIMATE CHANGE CLIMBS THE MOUNTAIN

Low temperature is one of the defining characteristics of high-altitude extreme environments. Climate warming makes these environments more hospitable to creatures from lower altitudes. From Labrador to Costa Rica, species have been observed expanding their ranges to places that were previously too cold or extreme. Warming allows plants and animals alike to advance to higher elevations. Tree lines rise as seeds take root farther up the mountainside. This expansion threatens the diverse and unique ecosystems that have evolved at the tops of mountains. Whether the mountaintop environment is a cloud forest or alpine tundra, the creatures that live there have no way to expand their range if plants from below move into their territory.

Antarctica: Frozen Desert

Antarctica, a continent with an area larger than the United States, contains 70 percent of Earth's fresh water. Much of this water remains frozen year-round. Only approximately 2 to 4 percent of Antarctica's land surface is free of snow cover for even part of the year.

Much of what is exposed during the Antarctic summer is rocky shoreline. Only a few plants are able to survive on land. With little rain or flowing surface water, Antarctica is technically a desert, one of the driest on Earth. Most Antarctic living creatures, including all its large animals—such as penguins, seals, and birds—depend on the continent's coastal waters for their food.

LIFE IN STONE

Plants and the few tiny insects and worms living in Antarctica must adapt to both extreme cold and dryness. Also growing in Antarctica are lichen, liverwort, and fungi, mosses where glaciers melt, and algae.

The only two flowering plants in Antarctica— hair grass and pearlwort—live in the warmest parts, on the Antarctic Peninsula and islands stretching toward Patagonia, at the tip of South America.

Some Antarctic organisms have even colonized the tiny pores of rocks for protection from the elements. Large-grained sandstone lets in enough light and moisture during the warm season for photosynthesis to occur while layers of algae, fungi, and lichen grow, extremely slowly, inside the stones.

ENERGY EFFICIENCY

Antarctic animals' marine food web is relatively simple. From seals to squid and albatross to whales, animals eat each other less than might

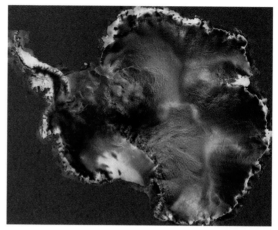

Top left: The moon rises near Davis Station, Antarctica.
Above: A satellite view of Antarctica.

be expected, and instead eat krill, small shrimplike creatures that are the staple of the local food web. Even crab-eating seals have teeth specially adapted for straining krill from the water.

The Antarctic marine ecosystem has high productivity in the warm season, when marine phytoplankton blooms support a dense population of krill. The amount of life supported by the Antarctic's coastal waters, from fish to land-dwelling animals, is surprisingly large. This is because the Antarctic ecosystem is so efficient: Many animals feed close to the bottom of the food chain, so not much energy is lost in layers of predation. The only top predators in the system are killer whales, birds called skuas,

The sea lion is one of Antarctica's few top predators. Krill is the staple of the Antarctic's simple food web.

and penguin-eating leopard seals. Because most other animals survive on krill, this challenging environment can support a large population of animals, including some as big as baleen whales.

LAND ANIMALS

The few year-round animals that get their food on land in Antarctica are all tiny creatures: a wingless fly, a mite, and a flea-like creature called a springtail. The seals and birds seen on the Antarctic ice, and even the penguins that spend months at a time there, get their food from the rich coastal waters.

Emperor penguins pay a high price for their inland breeding grounds. The penguins walk in single file up to 50 miles (80 km) to reach a spot where the ice is thick enough for their young to live on until maturity.

For as long as they are inland—the males incubating eggs and the females laying them and, later, caring for their chicks—emperor penguins depend entirely on food stored in their own bodies. Before they finish their tasks each breeding season and are able to return to coastal food sources, they may lose as much as 30 percent of their body weight.

For insulation, Antarctica's penguins, whales, and seals store blubber in thick layers. This adaptation seems quite simple compared to that of many Antarctic fish, whose blood is full of glycopeptides acting like antifreeze, allowing them to live in the icy waters.

Top: Penguins on a beach on Antarctica's coast. Emperor penguins prefer inland breeding grounds, then travel up to 50 miles (80 km) to reach coastal food sources at the end of the breeding season. Bottom: Few plants are able to grow in the frozen, dry environment of Antarctica.

Living on Ice

Few creatures can live their whole life on ice or snow, yet hundreds of species of algae thrive in Antarctica. Some are adapted to the extreme dryness of the interior; others can exist on the underside of sea ice, tolerating salt as well as freezing underwater conditions.

Algae on permanent snow and pack ice form the basis of food webs that include worms, fleas, mites, and other creatures in cold, nutrient-poor, and sun-blasted environments. Snow algae and ice algae are different from each other. Both can be red, green, or orange. A few species live in lichen communities.

Algae, which include organisms from single-celled phytoplankton (the basis of the marine food chain) to large seaweeds such as kelp, are among the first to colonize a new landscape and the last at the margins of where life can survive.

The creatures adapted to snow and ice environments can give scientists clues ranging from how to keep donated organs alive longer to what life might be like if it exists on other planets.

LAKES UNDER ICE

Miles underneath the ice in Antarctica lie liquid lakes. The largest of these is Lake Vostok, located under a glacier more

than 2 miles (3.2 km) thick. Lake Vostok was formed along geologic fault lines. It stays liquid because of the pressure of the glaciers above it and the heat from the geologic fault beneath. Lake Vostok is thought to have been sealed off from Earth's atmosphere for half a million years.

Samples of ice from near the surface of Lake Vostok—more than 2 miles (3.2 km) deep— contain microbes thought to have come from inside the lake. Scientists are working to create technology that will allow them to remove water samples from a subglacial lake and examine

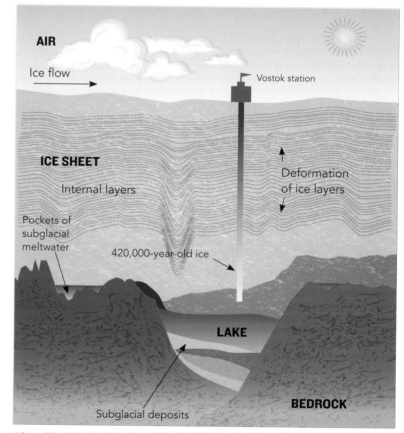

Above: The Lake Vostok system. Microbial life scientists have found bacteria, pollen, and marine diatoms in core samples taken from 420,000-year-old ice. Top left: A scientist walks near Lake Vostok Station, Antarctica.

them for ancient life in a way that does not introduce contamination from the modern world.

Lake Vostok is sometimes compared to Jupiter's moon Europa, which is thought to be made of ice. Since microbes have been found in ice cores thousands of yards below the Antarctic surface, scientists wonder whether life might also exist under Europa's ice.

METHANE ICE WORMS

Methane ice is a product of the high pressure and cold temperatures found on the deep ocean floor. Most methane ice lies underneath ocean sediment, but occasionally it seeps up and forms a frigid mound. It was at one of these mounds in 1997 that scientists first discovered methane ice worms.

In an environment without oxygen or sunlight—beyond the envelope of what had previously been considered necessary conditions for life—the one- to two-inch-long (2.5–5 cm) worms thrived. Methane ice worms have been found from Japan to the Gulf of Mexico. Living in small depressions like apartments in the methane ice, these newly discovered creatures are thought to feed on chemosynthetic algae and bacteria but may eat the methane itself.

GLACIER ICE WORMS

The glaciers of northwest North America, from Washington State to Alaska, offer a fairly harsh environment—not as extreme as Antarctica, but still not inviting to most life forms.

Yet billions of tiny worms inhabit these glaciers. At night they come to the surface to feed, mostly on algae. Some glaciers contain thousands of worms per square yard. The tiny brown-black worms, which resemble earthworms, are most lively and numerous at temperatures of just about freezing. Their metabolism speeds up as the temperature drops, though they cannot survive below 20°F (-7°C). And at temperatures above 40°F (4°C) their membranes start to disintegrate.

Top: Scientists near Lake Vostok Station. Bottom: Johns Hopkins glacier at Glacier Bay, Alaska. Billions of tiny worms inhabit the glaciers of northwest North America, from Washington State to Alaska. The small brownish worms resemble earthworms, emerging at night to feed on algae.

Hydrothermal Vents

Scientists used to think that the deep ocean floor supported little life, and in spite of repeated evidence to the contrary over the centuries, this myth persisted. Since sunlight can penetrate only to about 325 yards (almost 300 m), the top layer of the ocean was thought of as the "organic" layer, and below that, scientists believed, only scavengers could survive by eating morsels that fell down from above.

In the 1970s, when new technology allowed robots to explore the deep ocean floor, scientists were shocked to discover the teeming communities of life that were not connected in any way with the Sun's light.

The clouds of shrimp, giant spider crabs, and ethereal waving frills of giant tube worms in the white light of the robotic exploration vehicles could not be supported by organic material drifting down from above.

Deep-sea life is most dense around hydrothermal vent formations, where hot water gushes upward through chimney-shaped formations. The water is superheated in cracks between Earth's tectonic plates. Sometimes the water is black with hydrogen sulfide and other minerals, resembling smoke. The chimneys that spew this dark water are called "black smokers." Other chimneys are known as "white smokers" because they gush water whose chemistry and temperature color the water white.

VENT LIFE

The top predators in the hydrothermal vent community include several species of octopus, about a yard (1 m) long, that eat crabs, clams, and mussels. Another top predator is a two-foot-long (0.6 m) white eel-like fish.

Tube worms, up to four inches wide (10 cm) and two yards long (2 m), with red plumes that function as gills, are defining features of life in the vent ecosystem. Tube worms have no mouths or stomachs. Instead, the tubes contain huge numbers of bacteria living symbiotically with the worms, producing sugars from carbon dioxide and hydrogen sulfide in the vent, and oxygen from the surrounding water.

Above left: The giant tube worm, with its red gills. Above right: A hydrothermal vent has formed where the planet's crustal plates are slowly spreading apart and magma is welling up from below. Top left: Yellow iron oxide from a hydrothermal vent covers a rock formation.

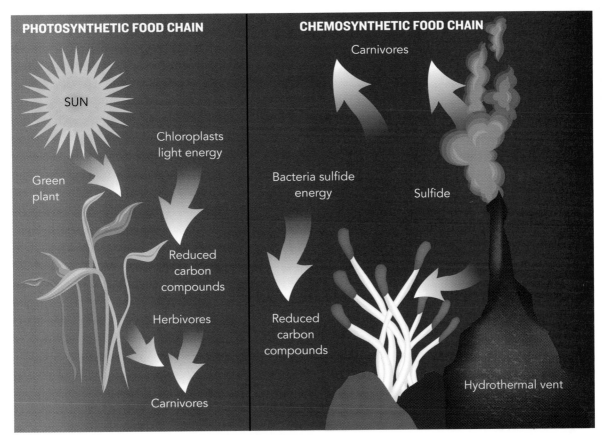

PHOTOSYNTHETIC FOOD CHAIN

SUN

Chloroplasts light energy

Green plant

Reduced carbon compounds

Herbivores

Carnivores

CHEMOSYNTHETIC FOOD CHAIN

Carnivores

Bacteria sulfide energy

Sulfide

Reduced carbon compounds

Hydrothermal vent

Organisms that exist near hydrothermal vents are not dependent on sunlight and photosynthesis, but instead rely on chemosynthesis, a process in which certain microbes use chemicals in the vent water to produce energy. They in turn form the base for an entire food chain of organisms.

Clams and mussels, like tube worms, depend on symbiotic bacteria that transform the chemicals in the fluid coming from the vent into sugars.

Several species of crab live near hydrothermal vents, filling niches from scavengers to fierce predators. Among other scavengers are colonies of jellyfishlike animals called dandelions.

Shrimp, feeding on microbes, swarm densely around the vents. Vent microbes include both bacteria and archaea, the most ancient form of life. The microbes transform carbon dioxide, chemicals, and energy from the vents into sugars through a process called chemosynthesis. Vent microbes form the basis of the community's food chain, similar to phytoplankton in the marine world above.

CHEMOSYNTHESIS

Before hydrothermal vents were discovered, having a source of energy from sunlight was thought to be a basic requirement for life. So when scientists discovered communities of creatures living without sunlight, their first question was, What is the source of energy in this ecosystem? The answer turned out to be primarily chemosynthesis. Chemosynthesis is the process by which microbes create organic compounds that can be used for food out of inorganic compounds. Chemosynthesis utilizes chemical energy to power the transformation. This stands in contrast to light, the energy source in photosynthesis.

Hydrogen sulfide, carbon dioxide, and methane are usually the raw materials of chemosynthesis. Many scientists believe that life originated in a low-oxygen, high-heat environment, so chemosynthesis may have powered the earliest forms of life.

Life Support

Life at Earth's extremes is a topic of ongoing fascination for astrobiologists, scientists who study how life might exist in outer space. The primary question for these scientists is, What are the basic conditions—chemicals, temperature, energy source—that are needed for life to exist? The amazing discovery by scientists of deep-sea communities that thrive without any connection to the Sun's energy sparked the imaginations of these scientists. Subsequent exploration in all corners of Earth has greatly increased our understanding of the types of environments that can successfully sustain life.

EXTREME MICROBES

Though we cannot see them, the most plentiful organisms around us are microbes, microscopic single-celled organisms that come in a variety of shapes and varieties. Microbes are the most common, sometimes the only, life forms in Earth's most extreme environments.

Ecologists are fascinated by the discovery that extreme environments can sustain life. They ask: What are the basic conditions needed for life to exist? Microbes have been found in such foreboding environments, as (clockwise from above left): an alkaline lake in Kenya's Rift Valley where the tiny organisms are the main food source for millions of flamingos; sea ice in Antarctica's McMurdo Sound; the high-salt waters of Israel's Dead Sea; and the center of the Atacama Desert, the driest place on Earth. Top left: Salt formation in the Dead Sea.

Researchers believe that primitive microbes are the life forms most likely to be discovered on other planets.

In recent years, microbes have been found in freezing, boiling, oxygen-free, and also extremely acidic environments. Until recently these places were considered unable to support life.

Microbes in alkaline environments of the soda lakes and alkaline hot springs of Kenya's Rift Valley are the main food source for millions of flamingos. These waters have an extremely alkaline pH level of 10, yet because of the microbes, the lakes are able to serve as breeding grounds for the flamingos, which both eat and provide nutrients to the microorganisms.

Once chemosynthesis was discovered, scientists became aware that all sorts of substances could be used as its basis. This greatly increased the number of places where life might exist. Microbes were found in sulfurous Japanese volcano systems, for instance; these organisms are thought to create energy through chemosynthesis, using sulfur as an energy source.

The Dead Sea got its name because scientists once believed it was devoid of life due to its high salt content. But new research shows that the Dead Sea is home to microbes called halobacteria that can live only in a high-salt environment. The family of salt-loving bacteria, called halophiles, is interesting because of the bacteria's expertise at repairing the damage that salt causes to their DNA.

MICROBES IN THE MANTLE

The hottest temperature at which a plant or insect can survive is about 122 (50°C). Some fungi can live at temperatures up to 144°F (62°C). But some single-celled organisms called archaea can live in temperatures up to 250°F (121°C).

Microbes have been discovered as far as down as 4.3 miles (7 km) below Earth's surface, embedded in the granite and basalt of mantle rock. These organisms are called endoliths, which means living inside rock.

The organic materials on which many endoliths survive contain the same carbon compounds that petroleum reserves are made of and were once on the surface but have been taken back into Earth's mantle.

Others live without oxygen and metabolize hydrogen. These microbes are thought to be related to the early microbes that lived when Earth's atmosphere was largely hydrogen, before photosynthesis began turning Earth into an oxygen-rich place that can support life as we know it.

STUDYING THE EXTREME

If extreme microbes' lives seem like a challenge, then studying

Old Faithful geyser at Yellowstone Park in Wyoming. Bacteria are now known to inhabit minute gaps in rock miles deep and live in geysers and boiling sulfur pools.

the biology and ecology of microbes that live in extreme environments is in itself a very challenging tasks. It is very difficult and expensive to re-create extreme environments in a laboratory, whether the goal is to cultivate these life forms for study or to re-create their challenges.

Scientists hope that by understanding how microbes survive in extreme conditions, how they repair their own DNA, or how they metabolize chemicals, they will be able to learn lessons that increase humans' ability to survive and adapt in difficult environments.

Extreme Diversity

Extreme environments such as the polar ice or the deep-sea bottom are difficult places to live. Species diversity in such places is very low, at least when it comes to plants and animals (scientists are just beginning to realize the extent of microbial diversity). Tropical rain forests are extreme environments at the other end of the spectrum.

The rich biodiversity of tropical rain forests is so rich that they harbor in excess of 40 percent of Earth's known species on less than 3 percent of the planet's landmass. More than 7,500 species of butterflies have been identified in Central and South America, compared with a mere 321 species in all Europe.

Why do some environments host such great diversity? The question of why biodiversity in tropical rain forests is so rich compared with biodiversity in much of the rest of the world is still actively being studied by ecologists. Several theories have been put forward.

PRODUCTIVITY

Tropical rain forests are a product of near-constant warmth and abundant sunlight and rainfall. Under these conditions, plants can grow efficiently year-round, creating lush, dense greenery and supporting a host of animals. Some scientists believe that tropical rain forests' biodiversity is the highest in the world because they were never covered by glaciers, because warmer weather speeds up physiology, or because without seasons the generational time spans are shorter.

Nutrients needed for productivity are recycled with unusual efficiency in rain forests by armies of detritivores including bacteria, fungi, and insects. The soil of tropical rain forests stores few nutrients; nearly all nutrients in the system cycle through its biomass at all times.

NICHE LIVING

Intense competition may lead to specialization and thus a greater number of species. In the highly productive tropical rain forest ecosystem, one species of bird may be dominant on just a single species of tree, or make its living in only the canopy, or very tops of trees. There may be room both for many predators because of the abundant prey and for many herbivores living on the constant foliage.

Competition in the tropical rain forest also may have encouraged extreme diversity by keeping any one species from becoming dominant. Tropical rain forest specialization has led to a high degree of complex symbiosis, which may allow for even more specialization to fill ever more niches.

Above: There is no shortage of plant life in the Carara Biological Reserve, Costa Rica. The rich biodiversity is a key feature of tropical rain forests. Top left: The blue morpho butterfly on a leaf. More than 7,500 species of butterflies have been identified in Central and South America.

CANOPY STRUCTURE

Though tropical rain forests are considered one ecosystem, they actually behave as different ecosystems layered one on top of another. As many as five levels of forest—canopy, overstory, understory, shrub, and ground level—harbor distinct communities, inhabited by insects, plants, birds, amphibians, and other organisms that live nowhere else.

The constant sun and high level of evapotranspiration—the rate at which trees and other plants lose water—along with plentiful rainfall, and the angle of sunlight at tropical latitudes, have all been offered as explanations for why rain forest canopy structure supports so many layers of rich diversity.

It is estimated that as many as 90 percent of species in tropical rain forests live in the canopy, which has not been thoroughly studied because it is largely inaccessible to ground-dwelling human researchers.

FOREST REFUGES

It has been suggested that during past ice ages, when sea levels were low and tropical areas became relatively dry, tropical rain forest habitats may have become separated for long periods into islands, or refuges. Geographic separation has been noted to facilitate the evolution of new species in other environments. With the levels of biodiversity already very high, these tropical rain forest refuges would have encouraged parallel bursts of growth in speciation.

As many as five levels of a tropical rain forest—canopy, overstory, understory, shrub, and ground level—harbor distinct communities, inhabited by insects, plants, birds, amphibians, and other organisms. Above left: A small parrot is perched on a branch in Brazil's Amazon tropical rain forest. Above right: Mosses and epiphytes on a tree in Isla Cocos, Colombia. Above: The rain forest brush on Isla Cocos, Colombia, covered with prop roots and epiphytes.

CHAPTER 10

THE VARIETY OF LIFE

Left: A grove of almond trees in full springtime bloom. The decline of the honeybee population in North America has California almond farmers searching for an alternative crop pollinator. A butterfly's iridescent wing (top) and some primates' taste for banana (bottom) are examples of how species and subspecies have developed special adaptations over more than a billion years of evolution. Increasingly, the fate of much of the world's diverse life lies in human hands.

During more than a billion years of evolution, life on Earth has evolved into a dense network, each species dependent on others, each evolved into subspecies that preserve special attributes—an iridescent wing, a taste for banana—and are adapted to local conditions. Species have knit into communities and ecosystems, covering the planet like a custom-made suit. Over the past 10,000 to 15,000 years, a blink in evolutionary time, humans have begun to play an increasingly powerful role in the state of many of the world's ecosystems. The fates of many species now depend on our choices. It may seem at times that humans are the only species that really matters in the world. Why should we care about having large numbers of other species around? Why care that 20,000 species of bee exist worldwide when only four species are raised for honey? The California almond farmer can answer, as he seeks an alternative almond crop pollinator after the decline of honeybees in North America. Both to humans, and to the Earth's ecosystems, the value of biodiversity, so long in the making, depends exactly on its fantastic variety.

Preserving Biodiversity

Biodiversity is the wide assortment of life on Earth. It is the genetic diversity within each species, the richness or total number of species on the planet, and the web of ecosystems through which life is woven.

Largely because of the rapid rise of human populations and their pressure on habitats and ecosystems, biodiversity is in danger at all these levels. Genetic diversity and ecosystems suffer as populations of wild animals and their habitats are divided by roads, farms, and towns into habitat islands.

Even though people are only one of 1.4 million known species, we consume 50 percent of the world's accessible fresh water and an estimated 20 percent of its plant growth annually. Because of human behaviors, more of those species are at risk of being lost now than at any time since the age of the dinosaurs.

"Species are going extinct at quite an alarming rate," says conservation biologist Stuart Pimm. "Routinely we have destroyed one natural resource after another." Pimm notes that in the past century, species have been going extinct at a rate 100 to 1,000 times faster than would be expected, considering fossil records. Other periods of rapid extinction have occurred in Earth's history, but if a mass extinction is indeed under way, this would be the first caused by a single species.

WHY DOES IT MATTER?

There are many ways of looking at why biodiversity is important. One is practical: Earth's ecosystems provide humans with essential services, from pest control to soil aeration, and biodiversity is key to those services, which would be impossible to replace.

A second is cautious: Genetic diversity provides backup so that life can survive the inevitable near-catastrophic floods, population dips, and other hardships that species experience over time.

Third is an aesthetic, or even spiritual, dimension: Earth would be a poorer place if geckos and jaguars disappeared, if polar bears no longer roamed the northern ice, California condors no longer nested in Big Sur, if the rain forests of Ecuador all became eroded cattle pasture. Perhaps the short answer is: All

Above: Las Vegas, Nevada, was a relatively unknown desert outpost during the first part of the twentieth century. In the past 20 years, the city's metro population has exploded to more than 1.5 million. Like many other population centers, Las Vegas's rapid growth is exerting great pressure on the area's biodiversity and natural resources, affecting ecosystems near and far as it seeks to satisfy its need for water. Right: The white rhino's origins can be traced back millions of years, yet in the last 15 to 20 years they have been hunted nearly to extinction. Top left: The Canadian Rockies provide a pristine environment for a number of species.

life is connected. The result of massive loss of diversity is difficult to foretell, and sure to be complex, but impossible to turn back. Following the last great extinction at the end of the age of dinosaurs, species did evolve and biodiversity returned—but the process took millions of years. While this might not matter in the grand geological scale, it might matter very much to the next few generations of humans.

HOT SPOTS

The distribution of species on Earth is quite uneven. Some places, such as the Arctic tundra, are native territory to very few species. Tropical rain forests—in Indonesia, the Amazon, or Hawaii—are home to the highest density of species.

Most land-dwelling species have small ranges (though a few common species, such as red-winged blackbirds, are exceptions). A great part of those land-dwelling species—around half of land plants and two fifths of land-dwelling vertebrates—are concentrated in small areas that together make up only a tiny portion of Earth's surface. These regions, called "biodiversity hot spots," are dotted around the world from Indonesia to Ecuador.

Hot spots used to constitute nearly 12 percent of Earth's land, but they have been logged and built on and degraded by human populations until they now constitute only 2.3 percent of Earth's land surface.

CONSERVATION STRATEGY

Preserving biodiversity hot spots is crucial to avoiding major species loss in coming decades. Poor people who live in or near endangered places must have a way to earn a living that does not involve degrading or destroying the land. Conservation money, rather than trying to keep people away from sensitive lands, can be spent on creating jobs and teaching methods of land management that help people to value their local ecology.

Meanwhile, people in the richest countries must change behaviors and policies that are causing long-term damage to Earth's climate and ecosystems. Sustaining the planet's life-giving ecosystems—which are threatened by construction, agriculture, deforestation, mining, and other activities of modern man—is as important as preserving species. Lowering carbon emissions by burning fewer and cleaner fossil fuels is just one place to start.

The mission of the nonprofit organization Conservation International is to "conserve the Earth's living natural heritage." The organization has characterized 34 locations around the world as biodiversity hot spots. These areas, shown on this map, hold a plurality of all plant and land vertebrate species over only a small fraction of the Earth's landforms.

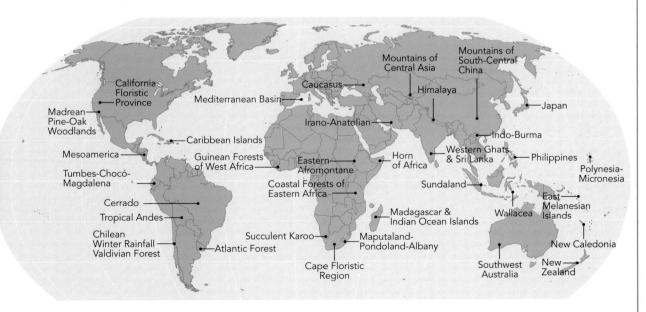

Stability and Resilience

Diversity of species may make an ecosystem more resilient to stress. Ecologists have observed that ecosystems composed of a large number of species seem more capable of recovering from disturbances than ecosystems with fewer species. This concept is similar to having a diversified investment portfolio. A portfolio spread among many stocks has far less risk of losing everything.

The reason for the correlation between resilience and diversity may be that a larger pool of species means more of those species eventually will find a way of becoming successful.

And if many different species are providing the same or similar basic functions within an ecosystem—protecting against erosion, for instance, or fixing nitrogen—then they can provide backup for one another if a disturbance occurs.

DISTURBANCE AND DIVERSITY

Every day a tree falls, opening a small hole in a forest. Every day a fire tears across a wood, clearing the way for new seedlings to find the Sun. Every day a landslide shears plants off a hillside. These disturbances violently intrude on old orders, creating opportunities for new species to move in and new relationships to develop.

Disturbance may be man-made, and it may occur slowly rather than abruptly. Its scale may be either benign or harmful. Overdrawing water from an aquifer can result in too much salt in the soil by drawing it in from surrounding soil and water. Agricultural runoff can shift the balance of nutrients downstream. Clearing a field and building a house can both change local ecosystems dramatically.

Natural or man-made disturbance, when it occurs on a moderate and not a drastic scale, can be an important part of ecosystem development and can promote greater diversity.

INVASION

Invasive species are non-native species that cause harm to the environments where they have been introduced. Invasives are a challenge to natural diversity all over the world today. As the Earth's ecosystems evolved over tens of millions of years, species reacted to, and developed relationships with, other local species. The population size and range of species depended on these relationships. Changes in temperature and rainfall, competition from other species, predators, and diseases are all threats that have elicited responses and adaptations.

Above: Wastewater treatment plants are built to help minimize the effects of man-made pollution. Still, the balance of nutrients downstream is easily shifted when some wastes, such as agricultural runoff, get into groundwater, rivers, and streams. Top left: A forest fire burns in the Canadian Rocky Mountains near Edmonton, Alberta. The resulting loss of trees will clear the way for new plants to take root and animals to move in. This sort of ecosystem disturbance contributes to an ecosystem's resilience and stability because the presence of a larger pool of species is known to help species become successful.

In an era of global travel, the previous balance of plants native to one location or another has been overtaken by a torrent of invasive plants and animals that, arriving in a new location, find no predators or less competition and are able to take over.

Invasive species alter local ecosystems in many ways. A field full of purple loosestrife, an invasive weed, is not as useful to marsh birds, migrating birds, or hay-eating animals as a field of native grasses. From Formosan termites to the kudzu vine, water hyacinth to starlings, invasive species are nearly everywhere. It is telling that more and more groups are organizing to promote "native plants," which now can be the ones that seem exotic.

CASCADING EFFECTS

Reducing biodiversity, even driving native species to extinction, is one of invasive species' legacies.

Invasions and disturbances are said to have cascading effects when they reach through the food chain. When ranchers in the midwestern United States exterminated most prairie dogs, they set off a cascade of effects in grassland ecology. Grasses changed when their soils were no longer aerated by prairie dog tunnels or their stems trimmed

by teeth; insects that used to live in the prairie-dog maintained grasses were unable to live in cattle pastures or were sprayed with pesticides; birds that used to live on the insects then departed.

Conservation biology is a field that seeks to preserve biodiversity where wild or native species have reached the point of requiring protection.

The Nile Perch is a voracious predator. In the 1960s, these fish were introduced into Africa's Lake Victoria in Tanzania as an experiment that resulted in the extinction of nearly all the lake's native fish species. The fish is now the major export of the region.

Formosan termites (above left) are considered an invasive species in the United States. These termites infest live plants and trees (above right). The U.S. Department of Agriculture started a campaign in order to reduce the termites' numbers and colonies.

Monoculture

The opposite of biodiversity is monoculture. Monoculture means growing or cultivating only one animal or plant variety at a time. An example of monoculture could be a cotton plantation, an orange grove, or a hog farm.

Monoculture has business benefits for farmers: It allows them to achieve "economies of scale." This means, for example, lower costs that come with large-scale planting, harvesting, storing, and selling a single crop. Because this type of farming resembles industrial manufacturing, it is sometimes called industrial agriculture.

But monoculture leaves no room for healthy ecosystems to thrive or to adapt when the going gets tough. Growing just one type of plant at a time, for example, makes that crop vulnerable to disease and pests. Industrial farmers often use a lot of pesticides to protect against this, but the pesticides kill off desirable insects, too.

On small farms that include both animals and plants, the animals' manure provides free fertilizer for the crops, and the animals can often eat some of the plants. But in large-scale livestock operations where only cattle or pigs are raised, the manure becomes a source of pollution as well as a costly nuisance.

POTATO FAMINE

One of the most famous monoculture disasters in history was the Irish Potato Famine. In September 1845, a fungus began attacking potatoes in Irish fields. Ireland's population, which had grown from 2 million to 8 million over the previous six decades, had come to depend on this one crop. Irish potatoes had little genetic diversity; none had resistance to the fungus. The potato crop quickly rotted.

Over the next four years, a quarter of all people living in Ireland starved or were forced

Left: When farmers grow the same plants year after year, their crops may be more susceptible to disease. Right: A potato infected with blight. A fungus was the cause of the Irish Potato Famine in the nineteenth century. The potatoes grown throughout Ireland were of the same type and had no resistance to the fungus. Top left: A farmer harvests his crop. Most large-scale farmers prefer to grow a single crop due to the economic benefits.

to leave their homes. A human society and its environment are closely intertwined. Ireland's potatoes were vulnerable to disease because potatoes were planted intensively to support a growing population on limited land. The people were vulnerable because of British policies, and because jobs at home had become scarce with the rise of industrial production in Britain. Many Irish people, like the potatoes themselves, were not able to adapt when blight hit.

FORESTS AND PAPER

Monoculture tree plantations are used by paper companies to raise trees that have been bred to fit the needs of the paper industry. A common plantation tree is the fast-growing loblolly pine.

Unfortunately, tree plantations are biologically poor places that do not fill the important ecological role of natural forests. They do not provide as much forage for grazing animals, or as much habitat for birds and small mammals. Creatures involved in breaking down and recycling dead plants, such as fungi, grubs, ants, and termites, are killed as the area is clear cut and burnt in preparation for planting. New growth of other plants and insects are inhibited through applications of herbicides and pesticides.

Because of the ecological value of natural forests, many scientists favor replacing tree monoculture with sustainable timber harvesting, a technique in which companies cut selected trees from natural forests while leaving the ecosystem intact.

Chinese farmers plant trees on a slope in Yanqing County, northwest of Beijing. Planting trees has helped to slow the erosion of outlying areas that has led to increasingly severe sandstorms in the capital city.

MONOCULTURE VERSUS BIODIVERSITY

Managing farms to preserve biodiversity can be even more profitable than monoculture. In southern China, a government initiative started a program to encourage farmers to plant trees in order to improve eroded land, with an eye toward harvesting timber when the trees matured in twelve years. Some farmers found that they could turn a profit in two years, rather than waiting twelve, if they planted mushrooms and native plants among the trees and raised chickens, vegetables, and bees for honey. Having a short-term income, the farmers were then able to plant valuable hardwoods, which would take longer to mature but earn more profit in the future. By thinking creatively, these farmers improved their quality of life and at the same time helped restore local biodiversity.

Fast-growing loblolly pines are the most widely cultivated timber species in the southern United States. Used for lumber and paper, these trees are grown on plantations, which do not sustain the diversity of flora and fauna that natural forests do.

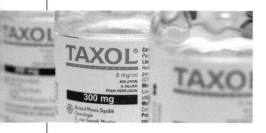

Biodiversity and Medicine

What cures for disease exist among the plants of the rain forest? Can these be discovered while there is still time, before the forest is cut down for grazing land? It is safe to assume that many plants exist, both in rain forests and other places, whose medicinal value is not known to scientists. According to the World Health Organization, a quarter of modern medicines are made from plants first used in traditional medicine.

Remedies developed from wild plants are used in the treatment of malaria, diabetes, cardiac illness, HIV/AIDS, cancer, pain treatments, and respiratory ailments.

The Pacific yew, which was once burned as useless garbage by western logging operations, was recently found to contain in its bark a substance called paclitaxol, which can help shrink cancerous tumors.

Some plants long recognized as having medicinal value have only recently been analyzed in a modern laboratory. Willow bark, for instance, was used for centuries to

Santalum album is an Indian plant that is more commonly known as sandalwood. Its widespread use in the manufacture of perfumes, incense, and traditional medicines has led to its placement on the endangered plants list.

relieve pain, but only in modern times was it discovered to contain salicylic acid, the active ingredient in aspirin.

ENDANGERED MEDICINES

Medicinal plants are known in countries from Brazil to Zambia, but China and India are two of the largest and oldest users of such plants. The WWF (World Wildlife Fund) has identified 33 herbs used in India's 3,000-year-old tradition of Ayurvedic medicine as being "critically

*Above: Pacific yew trees (*Taxus brefifolia*), the small evergreens in this photograph, grow in Oregon. A derivative of these trees produces the drug Taxol (top left), which is used in chemotherapy treatments of ovarian, breast, and certain types of lung cancer.*

endangered" and 17 more as "threatened." These include the elephant foot, red sandalwood, and lady's slipper orchid.

Medicinal herbs that have been used for thousands of years in the healing traditions of India and Tibet grow in habitats threatened by the expansion of human populations. Deforestation, urbanization, and overharvesting of wild plants endanger this biological wealth.

THREATENED HERBS
China and India are not the only places where traditional medicinal plants are being threatened.

The Ecuadorian plant cinchona, from which quinine, the first antimalarial drug, was derived, requires protection from overharvesting. The herb pheasant's eye (also called poet's narcissus), which grows in Eastern Europe and is used for heart problems, also must be protected.

Most medicinal plants are collected in the wild. This includes many commonly grown in the United States, including goldenseal, ginseng, echinacea, and American ginkgo.

Some medicines come from rain forests that are being cut for farmland. Medicine from tropical rain forests is not limited to plants; antibiotics and other medicines are derived from soil microorganisms, snake venom, and poison frogs.

As demand for successful herbal remedies grows, one way to improve conservation of medicinal plants is to increase cultivation in medicinal gardens.

SAVING PLANT LORE
In many landscapes that are in danger of being lost forever, plants have never been surveyed by ecologists or tested in modern laboratories for potential medicinal value. Holders of traditional plant lore are disappearing, and their store of centuries-old knowledge is going with them.

Organizations, including the World Health Organization and the National Institutes of Health, and botanical gardens have launched efforts to interview people who have traditional knowledge of herbs and medicines, to survey plants in rain forests and other wild places, and to transfer what is known about medicinal plants to modern medical laboratories that can study, record, and advance knowledge about possible remedies before they are lost.

A Hong Kong medicine shop sells traditional remedies. The history of Chinese medicine can be traced as far back as 1766 BCE.

Ecosystem Services

Even the most glamorous city-dweller in a fancy New York penthouse depends in the most fundamental ways on the services of nature. Most people spend their lives in man-made environments, and it is easy to forget the close ties to the natural world without which no man, plant, or animal would be able to survive.

People have come to rely on natural systems to create the oxygen we breathe and the water we drink; to purify pollutants in that air and water; to pollinate and provide pest control for crops, forests, and gardens; to break down garbage; and to produce the raw materials from which all products are made. Natural systems provide these services free of charge. Taken together, these gifts of nature that make life on Earth possible are known as ecosystem services.

PROVISIONING AND SUPPORT

The United Nations Millennium Ecosystem Assessment divides ecosystem services into four categories. One is provisioning services—products that humans obtain from the environment. This huge category includes food (crops, livestock, and fish), fibers for cloth, fuel, and fresh water.

The second category is regulating services. Natural systems maintain the balance between oxygen and carbon dioxide, keep Earth's temperature within livable boundaries, buffer against natural hazards such as flood, hurricanes, and erosion, and control the spread of disease.

Third are cultural services, defined as "nonmaterial benefits" such as spiritual enrichment, experience of beauty, inspiration, cultural heritage, chances for recreation, and the influence of environment on how societies develop.

The fourth category is support services, essential work that provides the foundation for these services. Support services include photosynthesis to support food production; chemical and biological processes for nutrient cycling; pollination and seed dispersal; and the work of decomposers to support soil formation.

Natural hazards like this flooding in Easton, Pennsylvania, cannot always be regulated by natural systems. Even man-made drainage systems are often overcome by heavy rains. Top left: Pollination and seed dispersal are categorized as ecosystem support services, as are photosynthesis and soil formation.

Although natural systems are resilient, they are not always able to bounce back from human practices such as overfishing.

PRESSURE ON THE SYSTEM

When people use river water to irrigate farmland, take fish from the ocean more rapidly than the fish can reproduce, or build houses where a wetland used to be, they are influencing ecosystems. Of course, not only humans can change ecosystems: When a beaver dams a pond, it is shaping that system as well. The services that the system can provide change along with it.

At any price, ecosystems cannot provide unlimited services. Though man is not the only creature to influence ecosystem services, the demands of our skyrocketing population on the planet's natural systems are testing the limits of those systems' abilities.

Natural systems are resilient; they can withstand many disturbances and bounce back. But the

The Croton Reservoir Dam was built between 1837 and 1842 as part of a system that was to supply water to New York City residents. The reservoir system was intended to meet the city's needs for centuries, but it quickly proved inadequate.

DRINKING WATER FOR NEW YORK CITY

New York City's water supply comes from rural watersheds upstate. By the 1990s, the quality of the city's drinking water was being affected by upstate building development and agricultural runoff. The city had two choices: It could build an artificial filtration system, or work to preserve the natural filtration taking place in the watersheds. Purchasing land, paying farmers to conserve sensitive property, and other programs to improve water quality through land management would cost an estimated total of $250 to $300 million. The cost of building a filtration system would be $2–$8 billion, plus millions in yearly maintenance. In this case, economics clearly illuminated the better ecological choice.

sharp decline of many fish populations in recent years, and signs that emissions of greenhouse gases are causing Earth's surface temperature to grow warmer, are two indications that we need to be very careful.

BIODIVERSITY AS A SERVICE

Biodiversity is one of nature's support services. Through biodiversity, natural systems grow stable, yet are resilient enough to withstand most disturbances that nature can throw at them,

from disease to ice to fire. The genetic and ecological diversity existing in Earth's environments today took billions of years of evolution to develop. Like other ecosystem services, biodiversity provides these enormous benefits free of charge.

It is up to humans to learn to perceive biodiversity not as just a fascinating display of nature's weird and wonderful imagination, but as the safety net that will keep our species alive and well through eons to come.

Values Beyond Value

Above: Poor farmers in countries like Costa Rica are generally more focused on their own survival than on the environment. Top left: Populations in wealthy nations like the United States consume more food, energy, and other resources than those who are less fortunate.

Would people value ecosystems more highly if dollar signs were attached? In an attempt to examine this question scientifically, economists assigned dollar valuations to services that are customarily given no specific economic equivalent. For example, economists have proposed paying the people who live near rain forests not to cut down the forests. How much is the service of the forest's biodiversity, and its service of removing carbon from the atmosphere, worth? The pay for conservation must be more than what farmers can earn from farming on clear-cut forest plots.

Economists reckoned that one hectare (roughly 2.5 acres) of former rain forest in Costa Rica could earn $125 per year as a farm, but the value of one hectare of preserved forest, which removes 7–20 tons of carbon from the air each year, should be $400. Unfortunately, this was only an academic exercise, as nobody is paying farmers to preserve the biodiversity and carbon sequestration services of the rain forest.

ENVIRONMENTAL JUSTICE

Ironically, both riches and poverty put pressure on the natural environment's ability to provide ecosystem services. Rich countries, with the United States in the lead, use far more services per person—more food, more energy, more water—than poor countries. If all people in the world used resources at the same level as Americans, there would not be enough resources to go around. Yet affluent people can afford to keep their immediate surroundings attractive, so that, to them at least, the pressure on natural systems is not obvious.

People in poor countries often feel they have no choice but to degrade their immediate environment in order to survive. Chopping down nearby forests for farmland and firewood, and killing wild, even endangered animals for sale or meat may obviously degrade their surroundings, but people who can find no other way of putting food on the table continue to do these things.

Even in the United States, poor neighborhoods bear a greater portion of the health and economic burdens brought by pollution and other environmental problems.

The struggle to get more money and attention spent on fixing these problems in poor neighborhoods is called the environmental justice movement.

STEWARDSHIP

The naturalist E. O. Wilson notes that every society possesses three types of wealth: economic, cultural, and biological. Economic wealth is the measure by which our societies are generally run. Cultural wealth is celebrated in many ways, through art and clothing and marriage rituals, for example. Biological wealth, if recognized at all, is treated as an insignificant part of the background.

It does not make sense, Wilson argues, to value only economic wealth. People should recognize that our lives, as well as those of our descendants and the lives of other species, depend directly on stewardship of the biological wealth of the land.

Some people believe that putting a dollar value on ecosystems misses the point, since our planet's life and landscapes can never be replaced and are therefore priceless. Instead of dollar signs, they argue, the concept of stewardship should be our guide. Stewardship means taking care of Earth for future generations. Stewardship implies an understanding that each of us is only a temporary passenger on planet Earth, and that we have a responsibility, as the Iroquois constitution said, to be mindful of the effects of our actions on future generations.

An almond ripens on a branch. Almond farmers in California have had to pay an estimated $100 million per year to maintain pollination of their crops, something they once took for granted.

POLLINATING ALMONDS

Like many things that are free, the tremendous benefits of ecosystem services often are taken for granted and undervalued by those who depend on them. Almond growers in California once could count on wild honeybees to pollinate their crops, but populations of honeybees have declined primarily due to disease. Honeybees, which are not native to North America, have played a role in reducing the population of native pollinators, along with habitat loss and pesticide use. This means that as honeybee populations decline, there are not as many native pollinators to take their place. California almond growers now must pay $100 million per year to rent hives to pollinate the almond crop. Throughout the United States, it is estimated that $5.7 to $8.3 billion worth of crops are lost due to declining pollinator populations.

Conservation experts with the U.S. Department of Agriculture survey the water quality in a mine water filtration pond in Somerset County, Pennsylvania. The project is part of federal efforts to protect the nation's resources.

Phlyogeny and the Tree of Life

The broadest definition of biological diversity, or biodiversity, is "the variety of life on Earth." But there are many different ways of describing this variety. One is on the level of genetic diversity—how many different versions of the same gene are there within a species? Genetic diversity is an important safety policy, ensuring that not all members of a species perish, even in the face of tough challenges such as disease or drought.

A second level is species diversity, the total number of different species on the planet. If you think of African wildlife as giraffe, elephant, wildebeest, zebra, termite, hornbill, and the baobab tree, you are thinking of species diversity.

A third way of looking at biodiversity is by way of ecosystem diversity. Without functioning ecosystem processes, life on Earth cannot survive. These include nonliving elements such as soil, water, nutrients, and climate, as well as living elements, comprising individuals, communities, and populations of different species that depend on one another for survival.

CLASSIFYING LIFE

Biologists classify life according to a system created by eighteenth-century Swedish naturalist Carolus Linnaeus. Species is the final unit of the Linnaean classification system. The classification of species, called taxonomy, builds a hierarchy whose largest category is a domain. The domain Eukaryota includes all life forms whose cells include nuclei inside membranes, which means all higher life forms.

Traditionally, species have been classified according to their phylogeny—their relationship through a common evolutionary ancestor. It can be difficult even for experts to determine scientifically how the branches of an ancestral family tree evolved. The study of how and through what criteria classifications should be determined is called systematics.

Here is the taxonomic classification of a bald eagle:

Domain: Eukaryota
Kingdom: Animalia
Phylum: Chordata
Class: Aves
Order: Falconiformes
Family: Accipitridae
Genus: *Haliaeetus*
Species: *H. leucocephalus*

Another way of classifying, called functional grouping, organizes life according to its place in the community or ecosystem. For instance, in a tidal community, functional groups might include algae,

Above: The brown pelican (Pelecanus occidentalis), like all members of the Pelecanidae family, has totipalmate feet. On each foot, all four toes, including the hind one, are connected by a web of skin. Top left: Tidepools are habitats that include a great variety of life.

bottom feeders, and swimming organisms, as well as creatures living in tide pools versus ocean. Functional groups might be studied for how they are distributed, or how they share resources or fit into ecosystem processes such as food webs and nutrient recycling. Functional groups may be organized according to their place in an ecosystem, either physically or based on behavior, habitat, or other characteristics.

Recent advances in technology, especially in the field of microbiology with its focus on the tiny, and genetic science with its advances in DNA sequencing techniques, have bolstered systematics and clarified family trees. These techniques have also led to the discovery of many unknown branches of the tree of life, even major branches where entire previously unknown phyla have been discovered.

Advances in classification in the 1980s led biologists to rearrange the basic view of the roots of the tree of life from five kingdoms into three domains: Archaea, Bacteria, and Eukaryota. Archaea and bacteria are both single-celled organisms that lack cell nuclei. Until recently they were lumped together as "prokaryotes." Improved understanding of archaeans' biochemistry and genetics made scientists realize that they were a completely different group.

BIODIVERSITY CONSERVATION

When deciding where to invest scarce resources, conservation groups have to decide what level of biodiversity is most worth addressing. The giant panda is the very last member of its genus, Ailuropoda. Does that make it more worth saving than, say, a whooping crane, which is only one member of the genus Grus?

Conservation may be focused on saving one species, or all the species in one particular place, or all that come together to make one ecosystem function. Classification, and the tree of life, are important ways to organize information and make these kinds of decisions meaningful.

THE TREE OF LIFE

All living things are connected by the passage of genes along the branches of the tree of life. Organisms sit like leaves on the outer edges of the tree. The history of their evolution can be represented by a series of ancestors, such as the single-celled archaea, primitive tubeworm, and primitive amphibian. These ancestors are shared hierarchically by different subsets of the organisms that are alive today, such as (clockwise on the illustration) rod bacteria, flowers, insects, humans, and frogs.

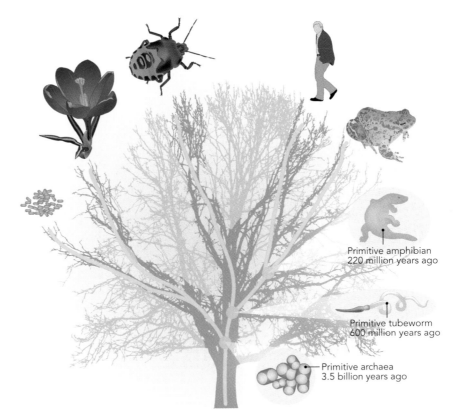

Primitive amphibian
220 million years ago

Primitive tubeworm
600 million years ago

Primitive archaea
3.5 billion years ago

163

The Unknown Sea

The basic concepts used to study relationships between living organisms and their living and nonliving environment— energy flows, population size and distribution, community functions, ecosystems—were developed for the study of ecology on land. When those same principles are applied to the sea, it becomes evident that marine ecology contains large knowledge gaps. It is safe to answer that many of the ocean's species, for instance, remain unidentified by scientists; life cycles and ecological needs also remain unknown.

FLOATING WORLD
Plankton are microscopic plants or animals that drift passively in water. Phytoplankton such as algae and diatoms produce energy through photosynthesis and are the primary producers of the ocean, supporting the vast majority of marine life.

Zooplankton are animals; they eat phytoplankton and are primary consumers in the ocean food chain. Some are larval stages of marine life such as corals and shellfish. All plankton play an important role in cycling carbon and other nutrients.

Only in the late 1970s did researchers identify the archaea, single-celled organisms that are one of the three domains of life

on Earth. Archaea are prokaryotes, meaning that like the earliest life on Earth, they have no cell nucleus.

Techniques from microbiology have recently allowed researchers to learn more about the microorganisms that dominate ocean life. In studying a very common member of the Roseobacter family, for example, scientists found that these bacteria were able to use sulfur and carbon monoxide to produce energy, in addition to organic carbon. This discovery has implications for the better

understanding of how nutrients, including sulfur, nitrogen, and carbon, are cycled in the ocean. The surprises revealed by these continuing discoveries only serves to underscore how much still remains unknown.

SPREADING PLATES
In recent years, ocean research has revealed astonishing facts about marine ecology. It was only in the 1950s that scientists discovered ridges in the middle of the Atlantic Ocean, confirming a basic fact of the nonliving ocean environment: that magma

These deep-sea vents are located more than 2 miles (3 km) beneath the surface. The ability of this region to sustain organisms such as tubeworms is something that scientists are only just beginning to understand. Top left: The movements of the water dictate the travels of plankton, which live at the surface of the sea and are weak swimmers.

from Earth's mantle constantly wells up between the harder crusts of continental plates. This discovery supported the theory of plate tectonics, the idea that Earth's crust is like a jigsaw puzzle of hard plates floating on top of molten rock.

Plate tectonics adds much to the knowledge of ocean ecology. For instance, where the continental plates come together, volcanic activity provides both heat and nutrients to deep ocean ecosystems. Because of the new material that comes out of Earth's mantle at the junctions of plates, and the old material that gets recycled back into the mantle, the ocean bottom is much younger than most continental rock.

ELUSIVE BOUNDARIES

Ecosystems, one of the basic units of ecological study on land, are difficult to define in the ocean, which lacks clear physical boundaries. This proves problematic in that only by understanding communities and ecosystems, the function of their parts and their role in relation to the environment as a whole, can good and well-thought-out decisions be made about managing the oceans, or keeping human activity from irreparably damaging them.

Better understanding of the species interactions, physical properties, and benefits to humans of the more localized systems, such as coastal seagrass marshes, coral reefs, and kelp forests, has helped guide environmental policies in a way that aims to balance human activities with the health of marine environments.

Above: Saltwater marshes are grasslands that are covered periodically by the rising tide. They are found where a river meets the ocean and are home to insects, crustaceans, and different types of grasses. Top: The diversity of life in a coral reef is said to compete with that in the rain forests.

CHAPTER 11

THE LEGACY OF MAN

Left: A farmer and his sons walk in the face of a dust storm in 1936 in Cimarron County, Oklahoma. Over-plowing and over-grazing that removed the native grasses from the Great Plains, combined with years of drought, turned much of the area into desert and produced dust storms that gave this region its name—The Dust Bowl. Top: By the end of the 1800s, logging operations were in full force in Washington State. Most of the logging industry used the method of clearcutting, wiping out large tracts of forest. This method led to the destruction of many forest lands. Bottom: Central Park in New York City got its start in 1858 when urban planners recognized the need for green space preservation.

Humans have transformed the planet. But so have bacteria and insects—we are not unique. If humans were still a relatively few 300 million tribal people as we were only 2,000 years ago, we would not be capable of using up the planet's resources, or otherwise disrupting its ability to support life. The amount of food, fuel, land, water, and other resources—some of them irreplaceable—that human's consume, is growing even faster than the population. Most people still do not realize that healthy ecosystems are critical for survival. One practical benefit of ecology is to help people understand both the limits of Earth's resources and how to keep from overstepping those limits.

Ecology—as distinct from environmental studies or conservation biology—is a "pure" science, focusing on understanding patterns and processes underlying the distribution and abundance of organisms. But as the human influence on the planet has grown, many people have combined the science of ecology with physical, social, and other sciences in an ethical framework, creating the applied fields of conservation biology and environmental science. The last two chapters of this book will focus on the broader field of environmental science.

Early Impacts

From the development of spears to stone and bronze tools, humans have used their technology and engineering skills to influence their environment.

Technological advances that have allowed humans to influence their world on a large scale include the development of agriculture and the domestication of animals. Social organization grew apace with settlement on the land.

Irrigation and the growth of cities fostered a more sophisticated concept of government, separation of skills, and larger-scale accumulation of material possessions. The conflict over control of resources and material possessions has begat an era of empire building. Mapmaking, boat-building, mathematics, and star-gazing have set humankind on the road to a future of manipulating the planet's delicate environment to its own advantage.

Human history is punctuated with stories of civilizations being disrupted by environmental crises. The abrupt end of Mayan civilization may have been brought on by extended drought. The Black Plague in Europe, which caused the death of 25 million people in the mid-fourteenth century, about one-third of the population, may have been exacerbated by preceding decades of climate-related crop failures and famine.

THE IMPACT OF GROWTH

Evidence indicates that the arrival of humans in new parts of the globe was followed by the extinction of many large mammals (also called megafauna—Latin for large animals) such as mastodons, mammoths, and saber-tooth tigers, ground sloths, stag moose, and giant beavers. Human hunting is one possible cause; other causes could be disease carried by early human migrants, or climate change.

Human impact on the environment broadened with the spread of agriculture. Large areas of land were cleared in Europe, Asia, and later North America. William Ruddiman, a climate scientist at the University of Virginia

Above left: Copernicus, a sixteenth-century astronomer, using a telescope. Advances in technology are one factor that led to humankind's great impact on the environment. Above right: The ruins of Chichén Itzá, a Mayan city in Mexico's Yucatán Peninsula. The downfall of the Mayan civilization may have been a result of extended drought in the region. Top left: Easter Island is located thousands of miles from the nearest population center. The first inhabitants were most likely Polynesians who arrived about 400 BCE.

in Charlottesville, connects the rise of carbon dioxide in the atmosphere 8,000 years ago to the development of agriculture.

Other scientists believe that the order was the other way around: Increased carbon dioxide released in the atmosphere by warming temperatures at the end of the last ice age allowed the evolution of cereals that became humans' first crops.

Whichever the order was, agriculture and domestication of animals provided a more reliable source of calories that allowed the human population to grow. They also led humans to place less value on wild plants and what we now call ecosystems.

LIMITS

As the population of humans grew, their impact on Earth's environment grew as well. From Easter Island to Iceland, land that was wooded before being settled by humans became virtually treeless as a result of house building, boat-building, and deforestation for agriculture, cooking, and heating fuel.

The population of humans was limited—by available calories, by the productivity of the land, by disease, and by the speed at which people and information could travel. Nevertheless, the population of humans and their impact on the planet continued to grow. Global trade brought invasive species, both plant and animal, that followed shipping routes. By 2,000 years ago, the human population had risen to an estimated 300 million. By the mid-eighteenth

CARBON DIOXIDE ON THE RISE

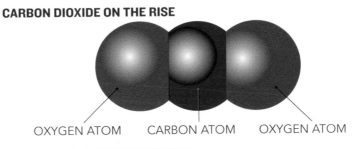

OXYGEN ATOM   CARBON ATOM   OXYGEN ATOM

CARBON DIOXIDE IN THE WORLD

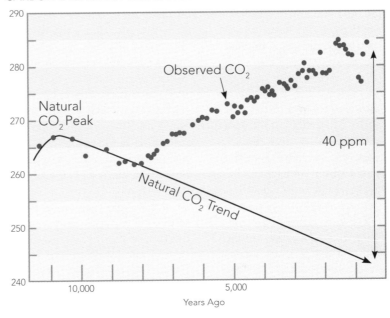

Years Ago

Above: This figure shows actual rise in carbon dioxide, measured in parts per million (ppm), over the last 8,000 years as compared to predictions of carbon dioxide trends without human intervention. Some scientists believe that humans caused the increase in carbon dioxide as a result of agricultural practices that included deforestation.

century, just before the start of the Industrial Revolution, the population of the world was probably nearly 800 million.

Right: Kudzu, which is native to Eastern Asia, was introduced to North America in 1876. During the 1930s, the U.S. Soil Conservation Service promoted the use of this highly invasive vine for erosion control. In 1972, the U.S. Department of Agriculture declared it a weed. Kudzu can envelop an entire tree and kill it by cutting out all light.

The Industrial Revolution

In the late 1700s and early 1800s the Industrial Revolution, spurred by the invention of the steam engine, led to the age of factories. The cities of England were transformed within a few decades, and the revolution quickly spread to the United States and the Continent. The Industrial Revolution was based on the idea that mechanical energy could replace human or animal labor. The source of that energy was steam, created by heat from burning wood or coal.

The burning of large amounts of coal caused smog in some cities that was said to darken the sky during the day like a blanket over the Sun. Factories poured soot into the air and dumped waste into the water. Sewage from increasingly crowded urban areas caused a stench to hang over the rivers. Visibility in the Thames was said to be no more than a few inches, the length of a calling card.

On the farm, machinery replaced manual labor for plowing and threshing. Garden plots and small, diverse farms could not efficiently be worked with these labor-saving devices. It became more economical to farm large parcels of single-crop land with machinery.

In the United States, the fertility of small and hilly New England plots on poor soil—cleared when the first settlers arrived—was declining just as industrial farm equipment was coming into use. Railroads extending rapidly in the middle of the nineteenth century, opening food markets to crops grown in more distant fields. Industrial farming practices and fossil fuel–driven transportation allowed the center of American agriculture to move to the flat and fertile Midwestern states.

LIVING CONDITIONS

The wretchedness of poor city dwellers, whose misery was captured most famously in the novels of Charles Dickens, was reflected in their often-squalid environment.

Workers migrated in great numbers to English slums as machinery devalued manual labor, forcing jobs that had previously been done in rural settings, like weaving and sewing, into factories. Factory workers lived in densely packed tenement housing in neighborhoods near the factories. The air in these neighborhoods was filled with the black soot and ash of the factory smokestacks.

Left: A polluted scene in Sheffield, England, in the 1800s. By the mid-nineteenth century, Steel City, as it was known, supplied nearly half the European output of steel. Top left: The harnessing of steam led to the development of powered machinery and the Industrial Revolution.

Garbage disposal was accomplished by throwing waste into alleys, to be cleaned up by free-roaming pigs. Dozens of cart horses died on the street each day and were left to be picked up by collection brigades. Conditions that were bad for humans proved good for disease-causing organisms. So many humans living in close quarters, with poor nutrition, bad air quality, and terrible sanitation, provided a nurturing environment for populations of several disease-causing microbes.

Epidemics of typhoid and cholera broke out regularly. More than 31,000 people died during a cholera outbreak in Britain in 1832. Cholera and typhoid, as well as dysentery, lung diseases, and tuberculosis, were understood to be connected with unclean water, pollution, overcrowding, and lack of sanitation, but germs and disease were still poorly understood.

AFTER THE REVOLUTION

Industrial production made many goods affordable to people who previously would not have been able to own them, things such as books, household appliances, and, later, vehicles. It greatly expanded the middle class, and accompanying scientific advances in hygiene and medicine eventually increased the health and life expectancy of the majority of people on Earth.

Mass-produced techniques and machines have grown much more efficient since the beginning of the Industrial Revolution. Cleaner energy production and manufacturing technologies have been developed. Laws and regulations have been passed since the mid-1800s to control the most obvious negative effects of factory production on people and land.

Still, many things have not changed since the Industrial Revolution. Poor people still often live in unhealthful conditions, and damage to the planet's ecosystems continues, caused by rampant population growth, harmful waste, overuse of nonrenewable resources, and a host of other reasons.

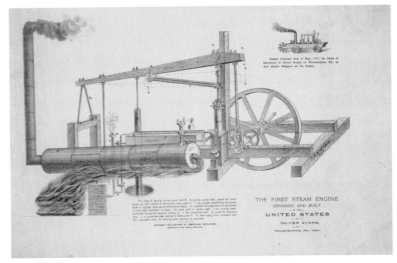

Above: The first steam engine designed and built in the United States. Oliver Evans invented this high-pressure engine, which advanced the milling industry by automating flour mills.

This mother and children in their apartment on Mulberry Street in New York City are fairly typical examples of the working urban poor in the early twentieth century.

Urban Growth

More than half the world's population now lives in urban centers. Worldwide, cities are growing both in population and in physical size. The United Nations forecasts that at least 12 cities will have more than 20 million inhabitants by 2025. The fastest growing cities are in the world's poorest countries.

The study of the ecology of cities is concerned both with the effects that cities have on the ecosystems and biodiversity around them, and how those ecosystems affect the cities and their inhabitants. Understanding how the ecology of cities works can both help avoid the burden cities place on their surrounding environment—local and global—and also improve the health and well-being of city dwellers.

DEADLY RESULTS

Cities' unhealthy interactions with their environment can cause polluted waterways, polluted air, and conditions that encourage outbreaks of infectious diseases.

Poor understanding of the physical world can be deadly. In Caracas, Venezuela, poor people built unplanned slums on steep hillsides, where construction and deforestation loosened the topsoil. In 1999, mud slides killed more than 15,000 people in these neighborhoods.

Above: Mexico City is one of the most highly populated urban areas in the world, with over 18 million inhabitants. The city struggles with poverty, political corruption, and numerous ecological issues. Top left: Industrial pollution is a major cause for concern in urban areas.

In New Orleans, Louisiana, long-term dredging and channeling of the river and draining and subsidence of the land left a coastline without its natural buffers and a city that was below sea level. Natural ecosystem services regulating water flow were replaced with machines—levees and pumps—to keep water off the streets. This lack of regard for natural ecosystems left the city of New Orleans vulnerable to catastrophic flooding during Hurricane Katrina.

ECOSYSTEM PLAYERS

Many ecologists have focused their efforts on understanding wild lands, places far from human activities. Humans have often been considered more of a disturbance to natural systems than a participant in the global ecological order.

More recently, scientists have come to realize that cities now play major roles in Earth's ecology. One goal of urban ecology studies is to help cities become accountable for the resources they use. The rate at which water filters through an environment has a direct effect on the productivity of that environment. Rainfall is funneled through the hard, nonporous surfaces that dominate modern cities—paved streets, parking lots, and gutters—more quickly than through the soil and wetlands of natural environments that absorb and filter water.

This rapid cycling of water through cities has several effects. One is that oil, garbage, and sewage may be washed in high concentrations into nearby

Above: Breaches in the levee system following Hurricane Katrina caused further flooding of the Ninth Ward in New Orleans days after the storm hit. The city was compromised in part by disregard for natural ecosystem services that might have regulated water flow in this area of Louisiana. Top: Lack of knowledge about the stability of building sites led to the death of thousands as a result of mudslides in Caracas, Venezuela, in 1999.

waterways when it rains. In natural ecosystems, soil and plant life act as filters for toxins and slow the rate at which they pass through the system, which serves to dilute their concentrations.

Cities can put stress on area water resources. This is particularly true in dry places, such as Arizona and Southern California, where much water is drawn from already stressed surrounding lands. In many cities, untreated sewage,

factory discharge, and other substances that are harmful in high concentrations are routinely dumped in waterways.

Increasingly, city planners are taking steps to filter rainwater and limit excessive water use. Encouraging the increased use of trees and foliage in medians, sidewalks, and parking lots is one solution. Riverside parks and reclaimed industrial lots can help buffer fragile waterside locations.

The Human Footprint

A heterotroph is an organism that cannot manufacture its own food and is dependent on others for food. Cities are heterotrophic—the energy and resources they consume cannot be produced on the land they occupy but must come from other places.

The cotton in a Los Angeles sundress may have been grown in Egypt and sewn in China. The fruit we eat in winter is likely to come from another hemisphere. The metal and plastic and concrete, the oil used to heat buildings, were all extracted from nature all over the world before being transformed into products.

An ecological footprint is a measurement of the total area needed to extract, produce, and dispose of the stuff consumed by people in a city, region, or country. Footprint data give a glimpse of the ecological impacts of consumption.

The footprint concept was developed by social scientists and economists in an attempt to show just how much of Earth's resources people, particularly in wealthy countries, use to maintain their modern lifestyles. The economists asked: If you count up the real cost of modern lifestyles, how long could this pattern of use continue?

OIL AND BEEF

It takes more than 300 gallons of water to grow a pound of rice, but 2,000 gallons to raise a pound of beef. If a family sitting down to a hamburger dinner never saw the cow, or the field where its grain grew or the truck that brought it to the city, the total ecological cost of that pound or so of beef remains invisible to them.

New Yorkers use about 1 billion barrels of petroleum per year, nearly half the oil produced each year in the United States. New Yorkers consume the equivalent to the wheat grown on over 1.9 million acres per year, about the size of the state of Nebraska.

Above left: The site of the former Blackbird Mine in Idaho. The site is the focus of efforts to remove soils that have been contaminated with arsenic and other toxic substances as a result of cobalt and copper mining operations. Above right: A cargo ship is loaded with grain at the Port of Seattle in Seattle, Washington. As the world's cities attempt to accommodate growing populations, they are finding it necessary to import energy and food resources. About 30 to 40 percent of the world's grain crop is fed to livestock. Top left: An open-pit copper mine. Once this type of mine has been stripped of its resources and closed, it must be rehabilitated to keep it from becoming a contaminated lake.

The ecological footprint of an average person in the United States is about 24 acres' worth of land, compared with less than 5 acres for a person in Brazil or Thailand. Americans consume more of everything: water, calories, oil, timber.

If everybody lived like Americans, we would need 2.8 planets' worth of resources to survive. This, of course, presents a problem. Fortunately for Americans, everybody does not consume as much, though people in developing countries do consume more each year in an effort to live more comfortable and healthier lives. Unfortunately for humans and other species as well, worldwide we are already consuming each year slightly more resources than Earth can produce.

WHAT TO DO

Even the apparent collapse of a once-thriving population, as happened in the North Atlantic cod fishery, is sometimes not enough to focus public attention. The footprint model provides one way for people in the United States and other wealthy countries to understand the consequences of their own lifestyles. The footprint model can be used to guide government incentives and actions by companies and individuals to encourage consumption levels that will be within the Earth's capability to sustain.

The lifestyles of people in wealthier nations create a larger ecological footprint and siphon a much greater share of the world's resources. Every year, New York City residents eat bread and pasta the equivalent of the amount of wheat grown on 1.9 million acres.

A hydrogen-powered Honda FCX. Environmentalists argue that while hydrogen power produces almost no pollution, it may lead to increased depletion of ozone. A possible compromise would be to add 5 percent hydrogen to gasoline, which could significantly lower emissions of nitrogen oxides and carbon.

WAYS TO REDUCE YOUR FOOTPRINT

- Eat food that is produced locally in season.
- Live simply, consume less.
- Replace some or all incandescent lightbulbs with compact fluorescent bulbs.
- Make sure the paper products you buy are recycled when possible.
- Make sure any cars or appliances you buy are energy efficient.
- Use curtains, fans, sweaters, and windows to regulate heat in the home.

Building Green

Ecologically conscious building and urban planning decisions and techniques can make cities healthier and more sustainable. These decisions begin with the knowledge that environments built by man depend on natural systems that make possible all life on Earth. The phrase "we are all connected" comes to mind not as a sentimental statement, but an ecological truism.

Green building describes construction and planning methods that are mindful of both where materials come from and the impact of a building or project on its environment after it is built. Green buildings use natural resources efficiently, getting more benefit per operating dollar or gallon of heating oil.

There are many products and techniques, from insulated glass to solar energy panels to natural ventilation systems that can make a building more resource efficient. In the United States, the federal government is heading an effort to outfit one million solar energy systems on buildings and homes by 2010. According to experts, maximum cost and energy savings are gained in a green building when the entire planning and design of a project, rather than just individual products, is undertaken with green goals in mind.

ECOLOGICAL BENEFITS

Green buildings use less fossil fuel for heating and cooling by employing techniques such as thicker insulation, and keeping out the Sun's heat in hot climates with special sun-filtering, insulated glass and supplementing fossil fuel energy with solar or geothermal energy.

Above: A "green" roof garden in Atlanta, Georgia. The city's chapter of the U.S. Green Building Council was formed in 2001 as part of an effort to reduce or eliminate the negative impact of buildings on the environment. Top left: A rooftop park six stories above the street in Tokyo, Japan. So-called green buildings are becoming popular around the world.

Cost advantages of green buildings continue throughout the life of the building. Green architects often take advantage of free ecosystem services, such as sunlight, wind, rainwater, and soil heat. Buildings from residences to offices and manufacturing facilities can be engineered to drastically cut utility and other operating costs, at the same time reducing carbon emissions into the atmosphere.

Green roofs, in which part or all of a roof is covered with plants, improve air quality, reduce heating and cooling needs, and filter rainwater runoff. Large, flat roofs are good candidates for greening. A roof covered with white, reflective roofing material instead of black tarpaper also helps save cooling costs and energy consumption in hot climates.

Landscaping with native plants that need less watering, collecting rainwater for irrigation, and controlling erosion during the construction process are some of the numerous other ideas that can help a building to be greener.

Since buildings consume about half of all the energy used in the United States, even relatively small changes in construction and operating techniques can add up to big environmental benefits, often translating to big cost savings as well. Organizations such as the U.S. Green Building Council, which oversees the Leadership in Energy and Environmental Design (LEED) rating system, helps builders and owners to plan for lessening the ecological impact of a building.

Above: The redevelopment of the Brewery Blocks complex in Portland, Oregon, incorporates ecology-friendly practices such as recycling of 90 percent of construction waste and solar technology. Top: NRG Systems, which develops wind power technology, incorporated solar panels and a cooling pond in the design of its new company headquarters.

WORKING GREEN

Occupants of a completed green building can increase its environmental benefits through many simple actions. These include turning off lights when no one is in a room, turning off computers and copy machines when they are not in use, and capturing the free heat provided by our bodies by keeping heat low while wearing an extra layer in cold weather. In warm weather, windows of green buildings can be opened instead of turning on air conditioning (in many traditional commercial buildings, this is not an option because windows do not open).

Recycling, both during and after construction, is a major source of energy savings. Building managers and occupants can buy recycled paper, and recycle paper, plastic, and electronic equipment.

Land Use

A parking lot built on a formerly wooded lot cannot filter air or water like the woods that used to occupy the land. Natural ecosystems provide many services related to climate, air quality, pest control, water flow, and the health of plants and animals. Expanding the built environment can threaten ecosystems' ability to perform those services.

As the population of humans in the world expands, the proportion of Earth's surface covered by vegetation—with all the services that natural systems provide—shrinks. The type of vegetation is also very important.

Land-use decisions are sometimes about buildings. Migrating birds that used to nest in a marsh will have to find new habitat when the marsh becomes apartments. But decisions may also be about converting land to agricultural use, constructing roads through forests, or clearing a stream bank to be planted with lawn. Land use is also an issue when ecologically important mangrove swamps are converted to tropical shrimp farming, as nearly 20 percent of them have been in Thailand.

When people build on land without consideration for natural ecosystems, they can cause harm to wildlife and even, ultimately, to human health. Since the

Above: Development on North Carolina's barrier islands poses a threat to surrounding estuaries and the nearby Croatan National Forest. Top: New agricultural settlements in deforested areas of the Bolivian tropical forest. Top left: A lakeside amusement area's sprawling parking lot extends to the water's edge, possibly having adverse effects on the local ecosystem.

Above: Many less affluent communities exist side-by-side with industrial plants. The residents of such neighborhoods are the focus of proponents for environmental justice.

loss of habitat is currently the biggest threat to biodiversity and is thought to be responsible for at least 80 percent of recent extinctions, change in land use is increasingly of great concern to ecologists.

DENSITY AND SPRAWL

Some ecologists believe that even though large, dense cities are a major intrusion on natural ecosystems, they make better use of resources than do low-density areas of sprawl. Though suburban areas have more trees and green space than cities, sprawl can be more damaging than intensive urban development for an ecosystem.

To fight sprawl, environmentalists encourage "smart growth," which means planning in ways that require fewer roads and encourage more walking and other efficient use of resources. For instance, planners of a

smart growth development in Sacramento built houses on smaller lots, within walking distance of stores, restaurants, and schools, surrounded by parks landscaped with native plants.

ENVIRONMENTAL JUSTICE

Often, the poorest members of a population live on the most polluted or undesirable piece of land in the community. Concentrations of industry, garbage processing facilities, pollution, and illegal dumping are more likely to be found near places where poor people live. Environmental laws may be enforced less rigorously in these neighborhoods.

The poor often suffer the largest health effects from land-use decisions—they are the ones with the most elevated asthma rates, the most hospital visits for heat prostration in hot weather,

the obesity in neighborhoods where there is no safe place to walk or play out of doors.

Environmental justice is concerned with fairness among people of differing economic status, race, and cultures. It is about how environmental harm and benefit are distributed among the wealthy and people who are poor or lack political power—about not having all the parks in a richer neighborhood and all the garbage transfer stations in a poorer one, with no safe place to play out of doors.

Poor communities in the American Southwest have asserted their right to control water use in places where the government sold public water rights to distant users, leaving local poor people, often including a high proportion of Hispanic and Native Americans, without enough water on their own land.

Above: An unofficial dumping site. Often, poor neighborhoods are strewn with such refuse where services such as garbage collection are not provided.

Nature and the City

Even in urban areas, natural systems continue providing the life-support services that keep ecosystems everywhere balanced so that life can exist. The soil, climate, and vegetation that support ecosystems are broken into patches in a city and its surrounding area.

Some of these patches will be very built-up, with little vegetation, providing few ecosystem services as the people in them greedily use resources. Other areas, particularly farther from the city center, may be rather rural and full of natural functions. This is true of virtually all cities.

In order to understand the ecosystem services that a city can provide, it is useful to take the whole metropolitan area into account. A study of the Baltimore metropolitan area uses the watershed as a basic unit, even though it reaches far beyond the city limits. Ecologists have studied the water, soil, vegetation, and biodiversity of the Baltimore area to gain an understanding of the energy and matter that go into supporting life in Baltimore.

Planted areas, in cities as elsewhere, absorb the Sun's rays and use a portion of the energy to create growth. Vegetation in urban ecosystems, in addition to helping to lower or even out temperatures, also processes water more judiciously, filters airborne pollutants, and breathes oxygen into the atmosphere. Ecologists can quantify these benefits.

Above left: The Baltimore metropolitan area from space. This false-color composite image shows vegetation as red, water as blue, and urban areas as grey. Urban sprawl surrounding Washington, D.C., and Baltimore has nearly merged them into a single megalopolis. Above right: The San Francisco Bay area is home to a variety of animal and plant species. Top left: Central Park in New York City provides an oasis for residents and wildlife alike. Approximately 26,000 trees and more than 275 species of birds can be found in the park's 843 acres.

SPECIES DIVERSITY

Diversity in cities includes more than the pigeons, rats, ants and cockroaches, molds and bacteria, dogs, cats, squirrels, raccoons, butterflies, and mosquitoes one would immediately think of.

Native plant species diversity is greater in cities with more variability—such as San Francisco, which includes redwood groves, coasts, and grassy hills—compared with diversity in cities that are more homogeneous, such as Chicago, which is located in the middle of a flat plain. Diversity of nonnative species is related to the age of the cities, or how much time the species have had to become established after being brought in by humans.

Port cities have more exotic, nonnative species, as would be expected of places where goods and people regularly are transported from other locations. Species diversity in cities follows some of the same rules it does elsewhere—diversity tends to be higher in warmer and more humid locations, higher when there are larger pieces of unbroken habitat, and lower at higher elevations and higher latitudes.

Cities generally have a relatively high component of nonnative, ornamental plant species, introduced by people, which may make the urban environment even less hospitable to native animal species than it otherwise would be. On the other hand, cities can encourage a resurgence of native bids and other wildlife by restoring landscapes using plants that are native and provide food or habitat to local wildlife.

WASTE

In a forest when a butterfly dies or a tree branch falls, the nutrients contained in that butterfly or branch are recycled with the help of detritivores such as bacteria, fungi, and termites. In natural ecosystems, nutrients are constantly recycled from the living to the nonliving environment.

In cities, large quantities of food waste, discarded packaging, and industrial waste are concentrated into landfills that sometimes are so unwelcoming even to detritivores that organic waste such as apple cores and half-eaten hot dogs can lie buried for decades without decomposing. Managing garbage is a challenge in cities in both wealthy and developing countries.

Above: Garbage is sorted in a municipal facility. Many cities in industrialized and developing nations target waste management systems with a focus on recycling. Still, the sheer volume of the waste to be managed is proving daunting to these municipalities. Top right: Fungus on this dying tree is an example of how natural ecosystems are able to efficiently recycle nature's by-products.

Burning Carbon

Carbon-based fuels have provided the energy that drove the rise of modern civilization. The use of carbon fuels—coal, oil, and natural gas—has been the basis of an astonishing rise in health and wealth for billions of people around the world. For people in less developed countries, the chance of a better lifestyle still is closely tied to burning increasing amounts of carbon fuel for electricity, cars, and to produce manufactured goods.

If fossil fuels had no dangerous environmental effects, humans only would have to worry about how long the fossil fuel supply can last. But burning of fossil fuels pollutes both the air and the water and has been linked with an alarming increase in Earth's surface temperature.

During the Industrial Revolution, the widespread burning of coal for steam power replaced animal labor as the major source of mechanical energy. Since then, atmospheric levels of carbon dioxide (CO_2) have risen by 30 percent. At about 370 parts per million, this level is higher than at any time in the past 650,000 years.

Other so-called greenhouse gases—including methane and nitrous oxide—have also risen. Greenhouse gases get their name because their presence in the atmosphere tends to hold in the Sun's heat, creating a warming effect like that of a greenhouse. An atmosphere kept within a certain temperature range is essential to life as we know it on Earth. But when the temperature goes above the limits to which life has adapted, it creates havoc in the planet's ecosystems.

Change in Earth's temperature caused by human activities is sometimes called "anthropogenic forcing." Since 1970, more CO_2 has been emitted from

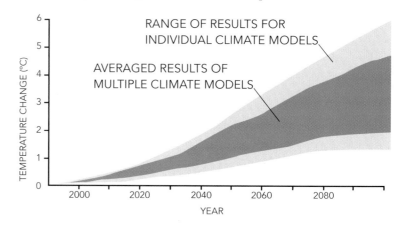

Top: Scientists predict that the global average surface temperature will increase between 2.5 and 10.1°F (1.4 and 5.6°C) over the next 100 years.

Above: Fossil fuel emissions have led to a change in global climate. The average temperature at Point Barrow, Alaska, rose more than 4°F (1.5°C) between 1971 and 2002. Top: Scientists predict that the global average surface temperature will increase between 2.5 and 10.1°F (1.4 and 5.6°C) over the next 100 years. The effect of humidity levels accounts for the wide variation in this projection. Top left: Oil refinery gas flares burn off waste gases, a likely source of greenhouse gas emissions.

Wind turbines convert wind energy into electricity. Wind power is one alternative to the burning of fossil fuels.

A No-Till drill is employed to plant soybeans. Planting field crops in this way, without plowing, may help to slow global climate change by trapping carbon in the soil rather than releasing carbon dioxide into the air.

CAPTURE AND STORAGE

While most efforts to reduce global warming have focused on limiting CO_2 emissions, some scientists have been working on ways to remove existing CO_2 from the atmosphere. Scientists say this is important because realistically, our civilization is going to continue to rely on carbon for the foreseeable future, no matter how dangerous it is for the planet.

There are two main ways of storing, or sequestering, carbon. One is by increasing natural carbon storage—planting more vegetation and not cutting down existing forests. The second is capturing CO_2 where it is created—at power plants and steel mills, and maybe, in the future, hydrogen plants—and pumping it into deep reservoirs in oceans or unused mines. New technology to capture carbon includes giant air filters to remove CO_2 directly from the atmosphere.

human sources than from natural ones. Scientists say the extent of recent warming can only be explained by human-generated carbon in the atmosphere.

GUARDING THE ATMOSPHERE

Scientists believe that in order to keep Earth's temperature from rising dangerously, CO_2 levels must be controlled. Climate systems are complex, and no one can say precisely how many CO_2 particles per million is too many. Climate scientist Wallace Broecker noted that letting carbon dioxide levels rise uncontrolled is like poking a sleeping dragon—very dangerous, even if you can't predict exactly how angry the beast will get.

Carbon in the atmosphere can be controlled in two ways. One is by putting less carbon into the atmosphere, either by increasing energy efficiency so that less energy will be used for the same work, or by using energy sources that do not involve burning carbon. The other is removing carbon that is already in the atmosphere. The current most practical way to reduce carbon emissions is using cleaner technology in power plants, cars, and trucks.

Better gas mileage is important. Increasing the amount of alternative energy, such as geothermal heat and wind power, will also help. Hydrogen fuel cells burn cleanly, but the technology for producing hydrogen still relies on fossil fuels. Conservation by individual people is important too. Turning off lights when you leave a room, using recycled paper, and riding a bicycle instead of driving all help to limit carbon emissions. But many environmentalists believe that only strong laws and incentives for power producers, vehicle manufacturers, and other key industries can cause a meaningful reduction in the amount of carbon in the atmosphere in coming decades and avert the worst problems of global warming.

The Cost of a Cell Phone

In the United States alone, over 100 million cell phones are thrown away each year. Cell phones are part of a growing mountain of electronic waste that includes televisions, computers, game systems, pagers, and personal digital assistants. The electronic waste stream is increasing three times faster than traditional garbage as a whole.

Electronic devices contain valuable metals, including gold, platinum, silver, and copper. A Swiss study reported that while the weight of electronic goods represented by precious metals was relatively small in comparison to total waste, the concentration of gold and other precious metals was higher in so-called e-waste than in naturally occurring mineral ore.

Electronic wastes also contain many poisonous metals, including lead, mercury, cadmium, and arsenic. Even when the machines are recycled and the toxic metals stripped out, the recycling process often is carried out in poor countries, in ways that are unregulated and allow many toxins to escape into the environment.

HAZARDOUS WASTE

The U.S. government's Environmental Protection Agency reports that more than 40 million tons of hazardous waste is produced in the United States each year. Some of this waste is as familiar as batteries and half-full paint cans.

Creating products out of raw materials creates much more waste material, up to 100 times more, than the material contained in the finished products. Consider again the cell phone, and imagine the mines that produced those metals, the refineries and factories needed to make the plastic casing, the box and packaging it came in. Many wastes produced in the manufacturing process are hazardous as well.

Some of the most hazardous wastes are those created by the nuclear power industry. Since high-level radioactive waste from nuclear power plants remains hazardous for hundreds of thousands of years, its disposal has been a long-term problem for the industry. In the United States not a single site has been approved for long-term disposal of high-level radioactive waste.

Above: Iron Mountain Mine in California produced iron, silver, gold, and other metals. Abandoned mines present daunting environmental problems to the areas surrounding them. Top: Computers are among the discarded electronic products that create what is referred to as e-waste. Top left: Cell phones contain metals and plastics that are refined and created in factories around the world. The manufacture of these devices creates hazardous waste as well.

The EPA notes that most waste is hazardous in that "the manufacture, distribution, and use of products—as well as management of the resulting waste—all result in greenhouse gas emissions." Individuals can reduce their contribution by creating less waste at the start—for instance, buying reusable products and recycling.

PRODUCER RESPONSIBILITY

In many countries the concept of extended producer responsibility is being considered or has been put in place as an incentive for reducing waste. If manufacturers are required to take back packaging they use to sell their products, would they reduce the packaging in the first place?

Governments' incentive to require producers to take responsibility for the packaging they produce is usually based on money. Why, they ask, should cities or towns be responsible for paying to dispose of the bubble wrap that encased your television?

From the governments' point of view, a primary goal of laws requiring extended producer responsibility is to transfer both the costs and the physical responsibility of waste management from the government and taxpayers back to the producers of the packaging.

Right: Yucca Mountain, Nevada, is being considered as a site for the first storage and disposal of high-level radioactive waste. Top: This chemical plant in Sillamae, Estonia, is only 110 miles from the nation's capital. This site provided enriched uranium to the Soviet Union for 50 years. When the Soviet military withdrew from the town, they left an estimated 12 million tons of radioactive waste.

CHAPTER 12

ECOLOGY AND THE FUTURE

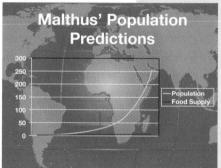

Malthus' Population Predictions

Left: Rainbow Bridge national monument, Utah. The bridge is sacred to Native Americans of the area, and visitors are asked to treat it with appropriate respect. This natural wonder is truly awe inspiring, as it stands taller than the U.S. Capitol building and reaches nearly as far as the length of a football field. Top: White Bengal tigers are an endangered species. Decreases in animal diversity have many negative effects, not the least of which is the loss of beautiful and irreplaceable flora and fauna. Bottom: This graph reflects Thomas Malthus's prediction that a population explosion would outstrip the world's food supply.

Ideally, people will figure out a way to utilize science and technology, combined with a respect for the Earth's life support systems, to engineer a safe future as human populations continue to rise. For this to succeed, we must understand those support systems and learn more about the planet's biodiversity and ecology, many aspects of which are still poorly understood. Ecologists say we know more about how many stars are in the universe than how many species live on Earth. There is a lot to learn.

Some things we know: As humans test limits of how far we can stretch water and land resources without creating environmental disasters, poor people will pay the heaviest price. We cannot know precisely where the dangers lie if we push the ecological envelope. Scientists' best recommendation is that people adopt a cautious, respectful approach to how the world's resources are used.

Ecologists like to point out that wilderness is the soul of human hope. It is the spirit of opportunity and freedom. What kind of a world do we want to live in and pass on to our children? It is to be hoped that the future world will be one in which this spirit survives.

Human Overpopulation

In the centuries since Thomas Malthus predicted that human population would outrun food supply, technology has moved faster than population. During the Green Revolution of the 1950s and '60s, agronomist Norman Borlaug and others helped increase the yield of food per acre by increasing irrigation, machinery use, fertilizer and pesticide use, and creating new varieties of wheat, maize, and rice that yielded much more food per acre.

The Green Revolution helped support growing populations in Mexico, India, China, and other countries and caused a major increase in the number of people the planet is able to feed.

According to population expert Joel Cohen, human population was about 600 million

Above: The metal plow is one of the advancements in farming technology that led to greater yields of food per acre. Top left: Despite its greater population density, the environmental impact caused by India's people is less per person than that of U.S. citizens.

in 1700 and more than tripled, to 2 billion, by 1927. In the half century between 1927 and 1974 the population roughly doubled again, to 4 billion. Enabled by the Green Revolution and medical advances, world population then increased

by yet another 2 billion, this time in only 25 years, between 1974 and 1999.

The success of the Green Revolution led some to question whether a carrying capacity existed at all for humans—perhaps technology would continue to allow us to exceed what were thought to be natural limits. But improved crop yields required industrial farming and heavy use of petroleum-based fertilizers. These intensive agriculture techniques temporarily overrode soil fertility limits and natural nutrient cycles but continue to have harmful effects on the environment.

As ecologist David Pimintel said, "Although improved technology can assist in more efficient production,

Higher yields of crops like corn and other grains come at a price. The use of petroleum-based fertilizers and heavy farming equipment can have detrimental effects on the environment.

The mid-twentieth century saw an increase in the use of fertilizers as well as the creation of new varieties of produce. Wealthy countries continue to benefit from increased food production.

The population of Shanghai, China, grew by nearly 3.5 million people between 1990 and 2000.

it will never be able to increase the supply of vital natural resources."

AGE STRUCTURE

The United Nations projects that the world's population will grow from its current level of about 6.3 billion to about 9 billion by 2050. The growth rate will be lower in rich countries. Birth rates tend to fall when people have better health care, lower infant mortality, are more educated, and benefit from economic stability.

Annual population growth is currently about 0.25 percent in rich countries and 1.45 percent in poor countries. The percentage of older people in rich countries is much higher than in poor countries, as a result of longer life spans and lower birth rates. The relative number of people in various age groups is called the "age structure" of a population.

Migration from poor to rich countries, which can affect the age structure, is more difficult to predict than birth rates. In some small, wealthy oil-exporting countries, migrants make up more than half the population.

ENVIRONMENTAL IMPACT

Joel Cohen notes that a population's environmental impact depends on many things other than natural resources, including how much energy and space each person consumes, social and cultural expectations, infrastructure for moving goods around, and government.

When considering how many humans is too many, experts look at the environmental impact of growing human populations. For instance, about one-third of the world's population, or 2.3 billion people, currently lives in China or India. People in China and India currently drive far fewer cars and use much less electricity per person than people in the United States. The population density of both China and India is greater, but the environmental impact of each person is smaller, than in the United States.

This pattern is changing as economies develop in both China and India and standards of living rise. Experts on global warming recognize that these two countries' increasing standards of living, translated into greenhouse gas emissions from cars and power plants, could dramatically increase the current global warming trend.

The Limits of Density

Slash-and-burn agriculture in tropical rain forests supported human populations for thousands of years. As long as the population density remained low, forests could recover.

Similarly, the Aral Sea in Central Asia and the two rivers that feed it had been used for irrigation for nearly 3,000 years. But with fewer than 8 million people living in the area at the beginning of the century, the population trying to support itself with irrigation did not exceed the available water supply. By 1997, more than 54 million people as well as large agricultural projects were trying to make a living off that same water—and the limits had been passed.

WHOLE-WORLD VIEW

As population expert Paul Ehrlich notes, if all the people in China and India lived in the continental United States, the U.S. would still have a smaller population density than England or Holland.

Population density is at one level a local problem. A growing human population that converts habitat to human use takes it away from the birds, amphibians, insects, animals, and plants that depended on that habitat, and it disrupts the food web connected to that land.

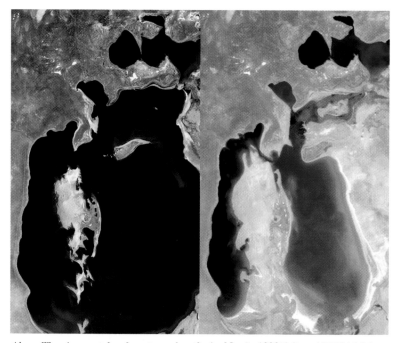

Above: These images, taken from space, show the Aral Sea in 1989 (left) and 2003 (right). Once the fourth largest lake on Earth, the Aral Sea has shrunk dramatically as a result of the diversion of the rivers that fed it. The water from the Amurdar'ya and Syrdar'ya rivers was diverted for irrigation of cotton farms and other agriculture. Top left: The growing human population can make it difficult for other animal species to survive in the same habitat.

Studies show that for a given area, as human populations grow they put pressure on other populations, from carnivores to herbivores to native plants. When does the habitat loss become deadly to other species?

Ehrlich points out that with only 55 people per square mile, Africa seems relatively underpopulated; in comparison, Europe (excluding Russia) has 261 people per square mile and

Japan a startling 857. Yet Africa still has depleted soil, large regions that have been turned into deserts, and forests that are being lost for firewood. In contrast, the majority of people in Japan and Europe are able to live well.

Global population density is not a question of how many people are crammed into any one small space, like commuters in a Tokyo subway at rush hour. To a point, having large

GLOBAL POPULATION DENSITY

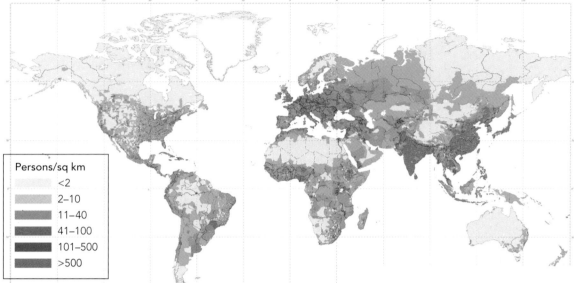

Persons/sq km
- <2
- 2–10
- 11–40
- 41–100
- 101–500
- >500

A bullet train at Tokyo Station. About 10 percent of Japan's total population (approximately 12 million people) lives in Tokyo, one of the most densely populated urban areas on Earth.

This map shows global population density in the last decade. Europe and Asia are the most densely populated continents on Earth. Having large numbers of people living closely together is not necessarily harmful to the environment, as long as enough open space remains to create a global balance.

numbers of people living close together in small spaces can be a good thing environmentally, leaving more open space outside the cities for other creatures to live in and for nature's life support systems to function.

The real issue with population density is on a planetary scale—whether the Earth has the resources to support a human population of 9 billion or more without causing mass extinctions and environmental disasters.

ROLE FOR ECOLOGISTS

Human population growth as a whole has not come up against the food constraints that Malthus predicted. Still, recent evidence of ecosystem stress (such as crashing frog and fish populations) and climate change (such as melting permafrost in the Arctic and rising sea levels) indicates people might be approaching another wall: Earth's limited resources.

Environmentalists have called this resource constraint "biophysical reality." The role for ecologists in helping prevent an overshoot catastrophe from wiping out large human populations is to continue to gather specific knowledge about Earth's ecology and how it reacts to human pressures, and to communicate with policymakers. Then, in theory at least, the policymakers will know what it means to make decisions that support ecosystems rather than destroying them, and correctly value the services they provide.

Poverty and the Environment

In Ethiopia, poverty means that farmers cut trees for firewood until there are no more trees left to cut, then burn dung for cooking instead of using it to fertilize fields, so that drought-stunted crops are even more meager, and the people live in a state of chronic malnutrition, becoming vulnerable to disease and death.

Each year millions of poor people around the world die of preventable diseases and malnutrition—they die because,

in the words of economist Jeffrey Sachs, "they are too poor to stay alive."

The actions of some of these people contribute to desertification and erosion and help create mud slides that kill thousands. It has been said that the poorest people on the planet—about 1 billion people live on less than one dollar a day—cannot afford the "luxury" of thinking about the future cost of degrading their immediate environment.

Many of the world's poorest people live in the tiny biodiversity hot spots where nearly a quarter of Earth's species are currently threatened with extinction.

THE COST OF DISASTER

A close-knit, two-way connection exists between natural disasters and poverty. Studies have shown that poor people are much more likely than rich ones to live on dangerous land such as floodplains, reclaimed wetlands, riverbanks, and steep mountain slopes. And the victims of disasters most frequently are poor people.

Natural disasters can set back economic development by years. Money that could have been used for economic advancement or environmental restoration is spent instead on emergency response.

Caught in a cycle of disaster and poverty, developing countries struggle to provide basic services such as education, medical care, and clean water. Economic development programs that might help people to earn a living in a way that does not damage the environment remain out of reach.

Even in the United States, natural disasters exact a greater toll on poor people, who, as in less developed countries, often

Above: This struggling farmer and father of eight lives in the lowland forests of West Africa, a biodiversity hot spot. More than a quarter of Africa's mammals, including chimpanzees, reside in this area where logging, mining, hunting, and human population growth are placing extreme stress on the forest. Top left: Ethiopia is one of Africa's poorest states. Political and economic instability have led to three decades of continuing dependence on food aid from international donors. Each year, regardless of harvests or rains, millions of Ethiopians need food aid for six months in order to survive.

live on marginal land, lack the means to get out of harm's way, and have few financial resources for recovery after disaster strikes.

In 2004, the Red Cross noted that environmental degradation and climate change are two of the major reasons why the cost of natural disasters is likely to continue to increase.

A GOOD INVESTMENT

Studies show that investing in the environment is a good way to fight poverty around the world. Environmental sustainability is an essential part of ending global poverty, according to the United Nations, which has made sustainability one of the top priorities for reducing global poverty.

Above: U.S. Department of Agriculture engineers inspect the soil of a dry parched stream channel.

Education, health, clean water, drip irrigation—investments in all these areas can help fight poverty in a way that reinforces the importance of healthy ecosystems. Investment in antipoverty development should involve local people who will be affected. Environmental restoration and preservation should be part of any economic development plan. Agricultural investment should support biodiversity and improved soil, not only new crop varieties. Investment in infrastructure, such as wells, communications, and roads, should support change that is ecologically sustainable.

Investment in new energy sources is essential for economic growth in developing countries, but carbon levels in the atmosphere must be controlled. A village that can cook with solar ovens will have no reason to cut down its surrounding forests for firewood.

Above: The World Summit on Sustainable Development was held in Johannesburg, South Africa, in 2002, only 25 miles (40 km) away from this community of Zevenfontein, the population of which is 85 percent HIV positive. Top: Costa Rican police officers evacuate a neighborhood in the small agricultural community of Alto Loaisa, where a mudslide buried thirteen houses and left six people missing in August of 2002.

Sustainable Development

Sustainable development means a chance to raise standards of living in the present, without destroying future generations' ability to meet their own needs. It contrasts with development that creates an impoverished future by using up nonrenewable resources (such as oil), and depleting renewable resources (such as trees and aquifers) faster than these can be replenished.

From an ecological standpoint, a population of any type of organism can survive over time only by living within its means. A population that overshoots its sustainability threshold—like the caribou on the island who reproduced faster than the lichen they fed on—increases the creature's risk of dying off.

Humans are not exempt from the principles of population biology. We are able to buy time by transporting goods from one place to another and modifying our daily environment, but we still depend on natural systems—photosynthesis, the water cycle, and so on—for survival.

Modern technology has sheltered us temporarily from the possibility that overshooting the planet's resource thresholds could lead to catastrophe. But science tells us that living sustainably will be a requirement of nature, and not a choice.

Top: Owens Valley in California, once a large saltwater lake, was dried up when the Los Angeles Aqueduct began diverting its water in 1913. Bottom: Mangrove forests that have withstood severe storms and changing tides for many millennia are now among the most threatened habitats in the world. In addition to environmental stresses such as pollutants, mangroves have been exploited by the shrimp farming industry, and are disappearing at an accelerating rate. Top left: Aggressive steps are being taken to improve the air quality in Los Angeles, which was ranked as the nation's most polluted city by the American Lung Association in 2004.

UNSUSTAINABILITY

A classic example of unsustainable development unfolded around the Aral Sea in Central Asia. The Aral Sea was one of the world's largest inland lakes. Starting around 1960, Soviet planners ordered huge quantities of water to be diverted for growing cotton, vegetables, fruit, and rice. Over four decades, the area's population grew from 7 million to 50 million, while the water-starved Aral Sea shrank by more than half.

The water became three times saltier than it had been previously. More than three-quarters of native fish species became extinct, wiping out a fishing industry that had supported 60,000 people. Salty dust on the shrunken shores was picked up by winds and dropped on fields, killing even irrigated crops.

The climate around the Aral Sea region has changed: Summers are hotter, winters colder, rainfall less. Crop yields have dropped, and drinking water supplies have been contaminated.

Health problems, including infant mortality and liver disease, hepatitis, and respiratory diseases, have skyrocketed.

Fortunately, when limits are recognized, not all environmental catastrophes are irreversible. Water levels in the Aral Sea have recently begun to rise again to levels that are viable for the four remaining species of native fish, thanks to an $86 million water quality restoration project.

EARTH CAPITAL

Economic development has often seemed at odds with environmental well-being. After all, belching black factories were the face of the Industrial Revolution. Wealth in modern times brings to mind SUVs and air-conditioned mansions more often than increased environmental consciousness.

Earth capital—the natural resources of the planet—is still undervalued and even considered without value in economic terms. Government subsidies are given for land-damaging activities such as industrial agriculture,

Owens Lake from space, February 2003. Once a large saltwater lake that covered hundreds of square miles, Owens Lake now contains a negligible amount of water. What little water remains is pink due to halophilic bacteria. The salt that now covers the basin is carried by winds to surrounding areas, damaging plant life and creating a health hazard for people in its path.

clear-cutting in old-growth forests, and grazing cattle on marginal federal lands.

Development decisions are made for economic reasons, though they may destroy this priceless capital. The United Nations conducted the Millennium Ecosystem Assessment to show that preserving the environment makes economic sense. The Assessment values tropical mangroves, when intact and acting as storm buffers, filters for pollution, and nurseries for marine life, at around $1,000 per hectare (2.5 acres). When cleared for shrimp farms, the former mangrove coast is worth only $200 per hectare.

A wetland in Canada, which also provides buffering, filtering, and fish-nursery services, had an estimated value of $6,000, while the same land used for intensive agriculture was worth only $3,000.

A wetland preserve in Canada provides a protected habitat for migratory birds.

Ecological Ethics

Ethics are moral standards that guide right and wrong conduct. An example of an ethical rule is the ecologist Aldo Leopold's dictum, "A thing is right when it tends to preserve the integrity, stability, and beauty of the biotic community. It is wrong when it tends otherwise."

Ethics are often framed in terms of rights or duties. Do people have a moral duty to conserve resources for future generations? Do animals have a right to be protected from extinction? Is it ethical to cut down trees or kill an endangered animal in order to avoid death? What if cutting trees will later bring death to others as healthy land turns to desert?

Differences over ecological ethics often can be reduced to the question of whether humans have greater rights than other living things. Belief systems can be roughly divided into three camps: 1) nature's ultimate purpose is to serve humans; 2) consideration of ecological costs and benefits must include all living things, not only humans; and 3) not only living things but Earth's ecological integrity—the ecosphere—must be included for any ethics considerations to be valid.

A person who takes the first human-centered approach might ask: How should Earth's resources be managed so that the planet can continue to sustain future generations of people? What is the minimum amount of wild land that must be set aside to avoid causing mass extinctions that could harm humans? Must all animals be saved, or are some not worth the cost? Ethics on this level can touch on local versus regional or even transnational issues. Does

American conservationist and ecologist Aldo Leopold pioneered the modern ecology movement with his book, A Sand County Almanac, which was published posthumously in 1949.

Above: In July 1980, the U.S. Congress set aside approximately 730,000 acres (295,421 hectares) for the Tetlin National Wildlife Refuge in Alaska. Some of the motives for creating the refuge were to conserve fish and wildlife populations and their habitats and to provide environmental education. Top left: A false-color image taken from space of croplands near Garden City, Kansas.

a town have the right to water its golf courses if that causes ecosystem distress downstream?

For the second, or "biocentric" camp, ethical questions grow broader: For instance, do people have an obligation to Earth's creatures, or only to other humans, to reduce carbon dioxide emissions if rising emission levels have been shown to cause environmental harm?

DEEP ECOLOGY

The ecosphere approach, sometimes called "deep ecology," holds that the well-being of the living and nonliving world has its own value apart from its usefulness to humans. Preserving the richness and diversity of Earth's ecosystems is of value, according to this approach, no matter whether that goal is beneficial to humans or not. Humans have no right to reduce diversity.

Deep ecology theory would contend that it would be ethically acceptable to reduce human populations in order to keep the rest of Earth's living and nonliving environment healthy, but it is unethical to sacrifice the health of Earth's ecosystems for the benefit of people.

ETHICS IN RESEARCH

The science of ecology makes advances through experiment and observation, which should be objective and therefore value-neutral. The results of scientific study, however, can have serious ethical implications.

For example, if stream-flow research shows that spring water peak is earlier by three days per decade over the past three decades, and if a mathematical model built to study the size and frequency of hurricanes shows that the storms are increasing as sea surface temperature warms, researchers are contributing to a body of evidence that global warming is occurring.

Is it ethical for government decision makers to avoid publicizing research that provides evidence of global warming? Or to shift funding away from field research on climate effects if the shift is motivated by a desire to downplay global warming?

Top: A Las Vegas golf course. Water conservation is an important issue for the inhabitants of this desert region. The Southern Nevada Water Authority maintains a water resource plan that includes suggested lawn alternatives for homeowners. Bottom: The Grand Canyon Wildlands Council takes a proactive approach to conservation. Its mission is to present positive, scientifically credible, and practical steps for redesigning how people live alongside nature. The members of the council hope that these collaborative efforts will help to sustain nature for many centuries to come.

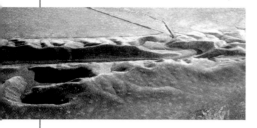

Hazards and Risks

In spite of a clear, undisputed record of serious earthquake hazards, people keep moving to California. And people in California continue to spend millions of dollars building on hillsides that have been recently washed away in landslides.

Hazards are dangers that could potentially cause harm. Risk is the chance that these hazards will in fact harm us. People often modify their behavior based on perceived risk, even though, when scientifically studied, perceived risk may be quite different from actual risk.

The chance that hazards will become disasters for humans —killing or seriously disrupting the lives of large numbers of people—depends on many human factors, such as whether those at risk have good early warning systems and strong houses, whether natural buffers such as marshes and mangroves have been allowed to remain intact, and whether homes have been built in high-risk locations, such as on a steep slope or right on an ocean beach.

ENVIRONMENTAL HAZARDS

Three types of environmental hazards are chemical, physical, and biological. Examples of chemical hazards in the environment are synthetic estrogen compounds

used in water bottle plastic and in the resin lining of food cans, and steroids used as growth hormones in both chickens and beef cattle.

Research indicates that these and other endocrine disruptors may have already affected average puberty onset dates and sperm counts in Americans. The chemicals persist in the water supply as well, affecting wildlife, for instance causing female fish to grow male reproductive parts and have fewer babies.

Physical hazards include what are often called "natural disasters," such as earthquakes, hurricanes, floods, and volcanic eruptions.

Top: Some give greater consideration to a view or neighborhood status than to the viability of the site when purchasing homes. Bottom: The extent of damage caused by natural disasters like earthquakes depends, in part, on human factors such as whether homes have been built with the possibility of hazards in mind. Top Left: San Andreas Fault runs almost the entire length of the California coast. Many people still choose to settle in California, however, in spite of the risk of earthquakes.

Earthquakes and volcanoes are caused by movements in tectonic plates and flows in the Earth's rock mantle, while water disasters such as floods, droughts, and storms are related to climate.

Global warming's effects on the water cycle may be leading to an increase in water-related hazards, such as stronger hurricanes in the Caribbean Sea and longer droughts in the American West.

Biological hazards generally refer to disease-causing germs, ranging from the flu to malaria, the mosquito-borne disease that kills more than a million people each year. Biological environmental hazards also include bacteria in unsanitary water.

Studies have indicated that some diseases such as AIDS, West Nile virus, Lyme disease, and bird flu may have become more prevalent in humans as a result of human populations' expansion into habitats where the disease microbes were formerly hosted by nonhuman animals alone.

MINIMIZING RISK

Evolution has shaped organisms' adaptations to risk in myriad ways, from the speckled camouflage of seagull eggs to succulent plants' strategy of storing water that can be used during droughts.

Animals' and plants' innate adaptations to environmental risks are developed over an evolutionary time scale of many generations. Not all organisms may be able to adapt to rapid change, such as the rate at which human populations have expanded their ecological footprint in the last century.

Humans and many other animals may respond to risk based on experiences encountered in one lifetime, such as whether a parent had lung cancer, or a favorite river became visibly polluted, or a particular nesting site proved dangerous. Humans sometimes use technology to protect themselves from hazards, such as dams and levees against floods, vaccines against disease, or specially designed buildings against earthquakes.

As Hurricane Katrina demonstrated, however, risk reduction techniques are not perfect.

Minimizing risk often involves a cost. For instance, many people who own beachfront property enjoy their ocean view and would be reluctant to trade it for a lower-risk property with less attractive scenery. Many more people around the world, though, live in vulnerable, high-risk areas involuntarily and do not have the means to move.

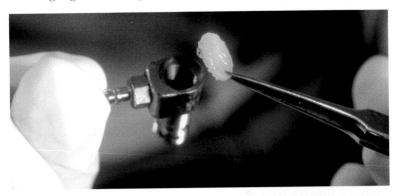

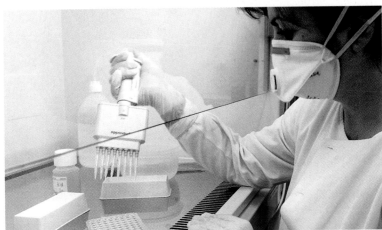

Top: A close-up of a material, or construct, that is a combination of living cells and porous biomaterial. This particular construct is implanted to enhance the repair and regeneration of bone. Technological advances like this one allow humans to protect themselves against diseases and health disorders. Bottom: A laboratory technician tests samples for Avian flu. Also known as bird flu, this strain of influenza spread from Asia to Europe through migratory birds. Creating vaccines to fight influenza viruses is an ongoing challenge for health agencies worldwide because these viruses are highly resilient and adaptable.

Trade-offs and Time Horizons

Frogs around the world are quietly dying off in large numbers, probably because of a fungus, habitat loss, pollution, and global warming. Ducks struggle to find places to nest or to rest along their migration route amid shrunken fragments of habitat.

Meanwhile, top political leaders continue to see growth and technology, as opposed to environmentally conscious living, as the key to a successful future. Environmental regulations are still perceived by many as anti-business. The majority of Americans do not yet embrace the idea of trading current living standards for long-term environmental health, or turning away from a consumer culture.

Accepting tradeoffs and acknowledging limits to Earth's ability to process carbon emissions in particular are more widespread in Europe. Other countries have begun paying far greater attention to the evidence that global climate change is happening and needs to be addressed immediately.

The United States, as one of two large industrial countries declining to ratify the Kyoto Treaty to limit greenhouse gas emissions, is in the minority among large industrial economies. Long-term degradation

of the environment still seems abstract to many people, and subsidies in the billions of dollars are still paid to timber, oil, and mining companies, and to farmers who use environmentally destructive agricultural techniques on a large scale. The majority of economic incentives in America seem more aligned with resource use than with conservation.

NO TRADE-OFFS NEEDED

Dale Bryk of the Natural Resources Defense Council notes that many businesses and

Rusting scrap metal turns a scenic Alaskan meadow into an impromptu junkyard. For many Americans, respect for the environment remains an abstract principle. Such attitudes lend themselves easily to long-term degradation of the environment.

Above: A sign urges U.S. President George Bush to sign the Kyoto Protocol in February 2005. The United States is one of only two industrialized nations that refused to ratify the Kyoto Treaty to limit greenhouse emissions worldwide. Top left: A satellite view of Planet Earth.

individuals are following another path, in which putting Earth's needs first seems less a sacrifice than an obvious choice—and perhaps even part of a successful marketing strategy.

Green buildings are showing economic as well as environmental superiority—filling up with tenants faster than traditional buildings and being less expensive to operate, while costing the same or less to build.

Eco-friendly businesses are building images and sales using the appeal of Earth friendliness. Strategies range from turning recycled soda bottles into fleece, to using organic ingredients, to emphasizing energy efficiency and recycling in industrial processes. These businesses often invite, rather than fight to avoid, environmental regulations, and even lobby Congress to pass carbon emission controls.

PRECAUTIONARY PRINCIPLE

There is no significant scientific debate over whether the globe is warming. It is. Five of Earth's hottest years on record occurred within the past decade. There is no serious scientific debate over whether negative impacts are already being felt in the environment. They are. Even if greenhouse gas emissions are stabilized today, the chain of effects that have already been set in motion will go on for centuries or more.

In the United States, while an increasing number of leaders on the state level have been mobilized into action, national political leaders have refrained from significant action. Many say that the specifics of Earth's response to climate warming and the relative contribution of human activities to that warming are not yet known, therefore we cannot as yet understand how much we should spend to lessen the risk from a future danger.

Nearly all scientists, on the other hand, agree that it is foolhardy to wait until lethal changes are clearly under way before taking action to prevent them. If we err, they say, we should err on the side of caution. Rather than looking at environmental cleanup or restoration costs as an unfair burden on people living today, we should be thinking of these costs as health insurance for future generations of Earth's residents.

Above: American bison are one of the success stories of species preservation efforts, having made a comeback in the last century from the point of extinction. Top: A typical animal waste lagoon in North Carolina. North Carolina's ten million hogs produce twice as much feces and urine as the populations of the cities of Los Angeles, New York, and Chicago combined. Industrial farms, most with thousands of hogs each, store the waste in the open-air pits. The farms produce more manure than the land can absorb.

GLOSSARY

A

ABIOTIC Nonliving, containing no living organisms. Abiotic influences on plants include temperature and rainfall (compare *biotic*).

ALTRICIAL Incapable of independent living at birth.

AUTOTROPH Organism able to produce its own food.

B

BENTHIC Living at the bottom of a body of water.

BIODIVERSITY Shortened term for "biological diversity." Can be measured in a variety of ways, including the number of species, the genetic variation within a species, or the variety of ecosystems.

BIOMASS Quantity of organic matter; total dry weight of all plants and animals in an ecosystem. Used as a measure of productivity.

BIOME A large geographic region characterized by a particular climate and dominant organisms.

BIOSPHERE All parts of Earth, including water, land, and atmosphere, that can sustain life.

BIOTIC Living (compare *abiotic*).

C

CARBON SINK A system with the capacity to accumulate or release carbon, that has more carbon flowing in than out during a given period of time.

CARRYING CAPACITY Maximum population of organisms that a given geographic area can support without being degraded or destroyed in the long run.

CHEMOSYNTHESIS Process of using chemicals to form organic nutrients from inorganic matter. An alternative form of primary production used by bacteria and archaea (compare *photosynthesis*).

CLIMAX ECOSYSTEM In the absence of any disturbance, the relatively stable, diverse, and productive state of an ecosystem reached after a succession sequence.

COEVOLUTION Evolution in which two or more species adapt to each other.

COMMUNITY All species living and interacting in a particular habitat.

COMPETITION Two or more organisms, populations, or species trying to live on the same resources in the same ecosystem.

CONSUMER Organism that cannot make its own food. A primary consumer (also known as an herbivore) is an organism that eats autotrophs, such as marine plants. Secondary consumers (also known as carnivores) eat organisms that eat other organisms.

CRYPTOBIOTIC SOIL Living soil crust, found in drylands and deserts, dominated by cyanobacteria and including lichen, green algae, moss, microfungi, and bacteria.

CYANOBACTERIA Prokaryotic, single-celled microorganisms in the domain Bacteria. Formerly called blue-green algae, though not algae. Among the earliest life forms to colonize land.

CYCLE A circular and continuous flow. Life on Earth is made possible by energy, water, and nutrient cycles.

D

DETRITIVORE An organism that feeds on detritus, or dead matter and waste from other living things.

DISPERSAL The movement of individuals to new locations, such as when offspring make a home away from their parents.

DISTRIBUTION The spatial range of a species in an ecosystem.

DIVERSITY The range of variation within a category, place, etc. In nature, diversity can occur at many biological, geological, chemical, and physical levels.

E

EARTH CAPITAL The natural resources and processes that sustain life on Earth.

ECOLOGY The study of how living things interact with each other and their nonliving environment; from the Greek "oikos," meaning "home," and "logy," meaning "knowledge."

ECOREGION A relatively large area of land or water with a distinct climate, environment, and assemblage of natural communities. Similar to biome.

ECOSYSTEM A localized system made up of living organisms interacting with each other and their nonliving environment.

ECOSYSTEM ENGINEER Organism that significantly transforms its environment, a prototypical example being the dam-building beaver.

ECOSYSTEM SERVICES Services provided for free by the Earth's natural systems, such as water filtration, crop pollination and irrigation, and oxygen to breathe.

EXPONENTIAL GROWTH Population growth pattern expected when no forces are holding back the population. When graphed, looks like a J-shaped curve.

ENDANGERED SPECIES Species in danger of becoming extinct due to diminished population, impoverished genetic diversity, or limited habitat.

ENDEMIC Native to a particular region.

EUKARYOTIC CELL Cell whose genetic material is contained in a membrane-bound compartment called a nucleus (compare *prokaryotic cell*).

EUTROPHICATION Processes by which bodies of water, such as lakes, receive extra nutrients that stimulate excessive plant growth and otherwise put the ecosystem out of its normal balance.

F

FITNESS In an evolutionary sense, the relative number of offspring produced by an individual relative to others belonging to the same species.

H

HAZARDS Potential dangers, ranging from earthquakes and hurricanes to ice on the front steps.

HETEROTROPH Organism that, unable to produce its own food, feeds on other organisms. See also *consumer.*

HYDROLOGIC CYCLE Ongoing, continual set of processes by which Earth's water passes through the land, sea, and atmosphere.

HYDROTHERMAL VENT Deep-sea formation in which superheated, chemical-rich water gushes upward through cracks in the seafloor, creating chimney-shaped formations.

I

INTERTIDAL ZONE Area between low- and high-tide water levels.

INVASIVE SPECIES Species capable of rapidly taking over a new area. Non-native invasive species are a major conservation concern.

K

KEYSTONE SPECIES Species that have a large effect on their ecosystem relative to the number of individuals present, such as prairie dogs and starfish.

M

MACROALGAE Algae growing in large, seaweed form, such as kelp.

MICROBE An organism so small it can only be seen using a microscope; also known as microorganism (compare *megafauna*).

MEGAFAUNA Large or relatively large animals (compare *microbe*).

MESOCOSM In ecology, medium-sized ecosystem simulation (e.g., terrarium, greenhouse) created to study ecological processes.

MICROCOSM In ecology, a small (less than the size of a mesocosm) simulation of an ecosystem, created to study ecological processes.

MONOCULTURE Form of agriculture in which only one crop is grown on a field at a time; from the Latin "monos," meaning "single," and "cultura," meaning "cultivation."

MORPHOLOGY Physical shape and structure of an organism.

MUTUALISM Relationship between two organisms in which each benefits from the other's presence, such as flowers and bees (see also *symbiosis*).

N

NATURAL SELECTION The process by which populations adapt to their environment as genes that increase an organism's probable reproductive success are preferentially passed on to future generations (see also *fitness*).

NICHE The place a particular species occupies within its ecosystem, especially regarding the food chain.

NUTRIENT Any molecule or substance needed by an organism to live and reproduce.

P

PALEOECOLOGY The science of reconstructing past ecological conditions using fossils, isotopes, chemical signatures, and other indicators.

PANGAEA The supercontinent that broke apart 200 million years ago to form all present-day continents.

PHEROMONE A chemical secreted by an individual that attracts, informs, or otherwise influences others of the same species.

PHOTOSYNTHESIS Process by which the light energy is used to convert inorganic material into organic nutrients (compare *chemosynthesis*).

PHYTOPLANKTON Microscopic drifting organisms in aquatic ecosystems, such as the ocean, that are capable of photosynthesis.

PIONEER ORGANISMS First living things to colonize new habitats, such as a new lava flow, which at the onset contains no life.

POPULATION Number of individual organisms of a single species living in a given area.

PRECOCIAL Capable of independent living at birth.

PREDATOR Organism that feeds by killing other organisms.

PROKARYOTIC CELL Cell with genetic material not contained in a membrane-bound compartment (compare *eukaryotic cell*).

PRIMARY PRODUCER Organism able to create food energy from light or chemicals (see also *autotroph*).

PRIMARY SUCCESSION The sequential process of change from one type of plant community to another, often more complex community in a place that has never before been occupied by living organisms.

PRODUCTIVITY The amount of the sun's energy converted to chemical energy by living organisms in a given place and period of time. Productivity usually corresponds to the amount of photosynthesis that occurred.

R

RENEWABLE RESOURCE A resource that can replenish itself faster than it is destroyed, such as forests that are harvested sustainably.

RIPARIAN ZONES The ecosystems surrounding streams; very important for wildlife.

RISK Chance or probability that a hazard will actually cause harm.

S

SECONDARY CONSUMER Organism that feeds on herbivores (see also *predator*).

SPECIES Often defined as a group of actually or potentially interbreeding organisms capable of producing fertile offspring, though this definition works poorly in many instances.

SECONDARY SUCCESSION The change over time of a sequence of communities reclaiming land after natural vegetation has been disrupted, such as by fire, farming, or development.

STRESS Anything that disturbs the normal functioning of an organism or ecosystem to the extent that its chances for survival are reduced.

SUSTAINABILITY The ability to consume resources without using them up. Economic development that takes full account of the environmental consequences of economic activity and is based on the use of replaceable or renewable resources that do not get depleted.

SYMBIOSIS Relationship between organisms living together in intimate contact. May benefit both or only one partner; in the latter case, it may be neutral or harmful to the second partner, as parasite and host.

T

TECTONIC PLATE Large sheet of rocky crust floating on top of the Earth's semifluid mantle.

TERTIARY CONSUMER Organism that feeds on carnivores.

TRANSECT A narrow strip along which researchers count organisms within communities to determine species' populations and variability.

TRANSPIRATION The process by which water evaporates from vegetation through small holes called stomata in its leaves or needles. Transpiration is one way water is cycled back into the atmosphere.

Z

ZOOPLANKTON Small or microscopic, drifting animals in aquatic ecosystems such as the ocean. May be larval stages of larger animals.

FURTHER READING

BOOKS

Carson, Rachel. *The Edge of the Sea*. Boston: Mariner Books, 1998.

———. *Silent Spring*. New York: Houghton Mifflin, 2002.

———. *Under the Sea Wind*. New York: Penguin, 1996.

Colinvaux, Paul. *Why Big Fierce Animals Are Rare: An Ecologist's Perspective*. Princeton, NJ: Princeton University Press, 1979.

Ehrlich, Paul. *The Machinery of Nature*. New York: Simon & Schuster, 1986.

Leopold, Aldo. *A Sand County Almanac.* New York: Oxford University Press, 1993.

Miller, G. Tyler Jr. *Living in the Environment*. Pacific Grove, CA: Brooks/Cole Publishing, 2000.

Ogum, Eugene, and Gary Barrett. *Fundamentals of Ecology.* New York: W. B. Saunders, 2004.

Wilson, Edward O. *The Diversity of Life*. Cambridge, MA: Harvard University Press, 1999.

WEB SITES

African Wildlife
www.wildwatch.com

Biodiversity and Human Health
www.ecology.org/biod/summary/index.html

Biomes
www.nceas.ucsb.edu/nceas-web/kids/
biomes_home.htm

Conservation Science
www.natureserve.org

Earth Science
science.nasa.gov/EarthScience.htm

Ecology for Gardeners
www.bbg.org/gar2/topics/ecology

Ecology of the Oceans
www.oceansatlas.org

Ecosystem Change
www.greenfacts.org/ecosystems/index.htm

Ecosystems
www.epa.gov/ebtpages/ecosystems.html

Extreme Environments
www.astrobiology.com/extreme.html

Fungal Biology
bugs.bio.usyd.edu.au/Mycology

Hydrothermal Vents
www.pbs.org/wgbh/nova/abyss/life/
extremes.html

Keystone Species
www.prairiedogs.org/keystone.html

Understanding Evolution
www.pbs.org/wgbh/evolution/library/index.html
evolution.berkeley.edu

MAGAZINES

Ecology & Society | www.ecologyandsociety.org

National Geographic | www.nationalgeographic.com

Natural History | www.naturalhistorymag.com

New Scientist | www.newscientist.com

Oceanus | www.whoi.edu/oceanus

OnEarth Magazine | www.nrdc.org/OnEarth

ScienceDaily | www.sciencedaily.com

Scientific American | www.sciam.com

Smithsonian | www.smithsonianmagazine.com

ORGANIZATIONS

American Museum of Natural History | www.amnh.org

Brooklyn Botanic Garden | www.bbg.org/gar2/topics/ecology

Center for International Earth Science Information Network | www.ciesin.org

The Ecological Society of America | www.esa.org

The Environmental Literacy Council | www.enviroliteracy.org

The Institute of Ecosystem Studies | www.ecostudies.org

Intergovernmental Panel on Climate Change | www.ipcc.ch

National Museum of Natural History, Smithsonian Institution | www.mnh.si.edu

Natural History Museum (London) | www.nhm.ac.uk

Natural Resources Defense Council | www.nrdc.org

The Nature Conservancy | www.nature.org

Sierra Club | www.sierraclub.org

Smithsonian National Zoological Park | nationalzoo.si.edu/ConservationAndScience

Union of Concerned Scientists | www.ucsusa.org

U.S. Environmental Protection Agency | www.epa.gov

U.S. Geological Survey | www.usgs.gov/science

Woods Hole Oceanographic Institution | www.whoi.edu/

World Resources Institute | earthtrends.wri.org

WWF International (formerly known as the World Wildlife Fund) | www.panda.org/

AT THE SMITHSONIAN

The Smithsonian Institution in Washington, D.C., is the world's largest museum and research complex. Composed of 19 museums and the National Zoo, the Smithsonian's exhibitions offer 24 million visitors per year a glimpse into its vast collection of over 142 million objects. The National Zoo and the National Museum of Natural History are among the planet's finest public resources available for furthering our understanding of the natural world.

THE NATIONAL ZOO

The National Zoological Park is a diverse organization, created to study, protect, and educate the public about wild animals and their habitats. Exhibits ranging from African Savannah to Backyard Biology enlighten the public about their effect on the environment and allow visitors to better understand the world around them.

Visitors watch the National Zoo's two adult giant pandas, Mei Xiang and Tian Tian, who arrived at the zoo on December 6, 2000. Their male cub Tai Shan was born July 9, 2005. American scientists, who are leaders in the field of giant panda biology and conservation, continue their research with these three pandas-in-residence as well as giant pandas living in the wild in China.

A view of the museum complex extending behind the Smithsonian Castle on the Mall in Washington, D.C.

In addition to the exhibits, the National Zoo has created educational and scientific programs, such as the Conservation and Research Center Program, both on-site and around the world. The strength of the Zoo's scientific research is in reproductive biology, veterinary medicine, behavior, conservation ecology and nutrition, population management, biodiversity monitoring, and professional training in these disciplines.

THE NATIONAL MUSEUM OF NATURAL HISTORY

The Natural History Museum is dedicated to understanding the natural world and our place in it. It is the largest of the Smithsonian museums, with 30 exhibit halls and a collection of specimens and artifacts that total more than 25 million—the largest, most comprehensive natural history collection in the world. The museum welcomes 6 million visitors annually from around the world.

The Museum's research scientists—nearly 400 biologists, geologists, and anthropologists—explore ocean depths, mountain peaks, rain forests, and locations across America to understand our planet and the forces that threaten its natural resources. In addition to the museums, the Smithsonian operates the Smithsonian Tropical Research Institute in Panama and the Smithsonian Environment Research Center in Maryland. Through these and other organizations, the Museum strives to further the dissemination of knowledge through promotion of science education programs.

SMITHSONIAN TROPICAL RESEARCH INSTITUTE (STRI)

The Smithsonian Tropical Research Institute (STRI) is the world's premier research institute for basic science in the tropics, dedicated to increasing our understanding of life in the tropics and its relevance to human welfare. Located in Panama, STRI is the only Smithsonian bureau based outside of the United States. The international

The National Museum of Natural History in Washington, D.C., houses the world's largest, most comprehensive natural history collection. Researchers at the museum investigate vital topics such as global warming, the loss of biodiversity, and invasive plant and insect species that threaten our agricultural and natural resources.

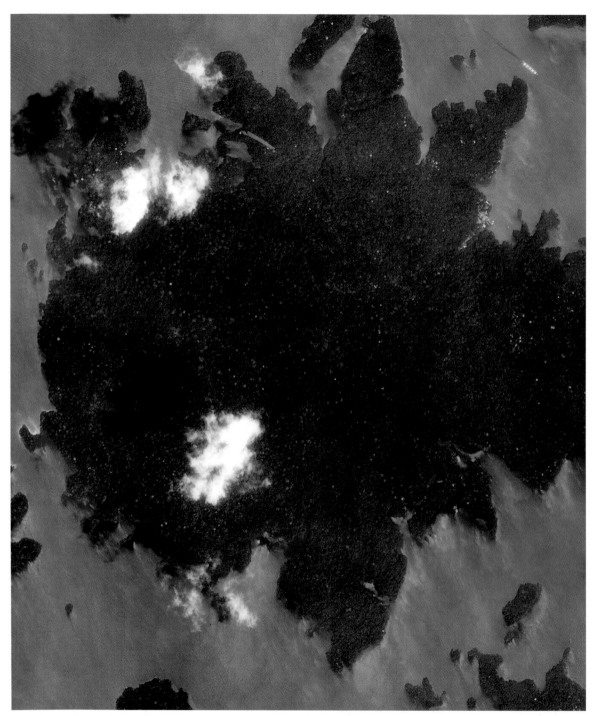

Barro Colorado Island is located in Gatun Lake, at the northern end of the Panama Canal. The red roofs of the Smithsonian Tropical Research Institute can be seen peeking through the dark green canopy inland of the bay on the northeast shore (top right in this photo). Barro Colorado Island has been managed by the Smithsonian since 1924 and is one of the premier sites in the world for the study of tropical forests and the plants and animals that live there.

staff of 40 scientists and approximately 500 visiting scientists and students each year concentrates on basic research in tropical forests and coral reefs, focusing on ecology and evolution.

The Institute shares their findings through a wide array of programs, including training opportunities for students worldwide, lecture series, supplementary textbooks, and hands-on learning experiences at their Marine and Terrestrial Exhibition Centers at various locations in and around the nation of Panama.

SMITHSONIAN ENVIRONMENTAL RESEARCH CENTER (SERC)

The Smithsonian Environmental Research Center (SERC), located on 2,800 acres along the shores of the Chesapeake Bay and the Rhode River, is the world's leading research center for environmental studies of the coastal zone. Some of the greatest environmental challenges lie in this zone, where 70 percent of the world's population lives. For over 40 years, the SERC has been involved in research, professional training, and environmental education. Research at the SERC addresses issues from global change to protection of fragile wetlands and woodlands. SERC researchers also conduct studies at field sites around the world—from Australia to Belize and Antarctica to Alaska.

The Institute's outreach programs have attracted a worldwide audience of up to 20 million through video conferences and electronic field trips. For those fortunate enough to live nearby, SERC offers hands-on science education programs, teacher-training workshops, and educational programs for adults.

OTHER RESOURCES

Numerous online exhibitions can be accessed via the Smithsonian Institution's Web site (www.si.edu). Presentations such as *Earth Today: A Digital View of Our Dynamic Planet*, and *Ocean Planet*, can be explored online. A broad range of articles related to ecology can be accessed on the *Smithsonian Magazine* Web site (www.smithsonianmag.com).

The Chesapeake Bay Bridge. The Smithsonian Environmental Research Center is located 25 miles from Washington, D.C., along the western shores of Chesapeake Bay, which is the largest estuary in the United States. Coastal environments are the focus of the center's research.

INDEX

ACKNOWLEDGMENTS

The author's deep gratitude goes to the fine and supportive team of editors at Hylas Publishing: Glenn Novak, Aaron Murray, Molly Morrison, and Roger Ochoa. I would like to thank my excellent science advisor, Jennifer Hoffman, for her steadfast assistance. The design staff at Hylas does a fantastic job. Also I would like to acknowledge all the inspiring scientists at the Earth Institute at Columbia University and the Lamont Doherty Earth Observatory, who have given me so much to think about. To my parents, who introduced me to both science and writing. And last, but certainly not least, thanks to Walker, Jack, and Leo for everything.

The publisher wishes to thank Gary Krupnick, Smithsonian National Museum of Natural History; J'Nie Woosley and Ann Batdorf of the Smithsonian National Zoological Park; Ellen Nanney, Senior Brand Manager with Smithsonian Business Ventures; Katie Mann with Smithsonian Business Ventures; Collins Reference executive editor Donna Sanzone, editor Lisa Hacken, and editorial assistant Stephanie Meyers; Hydra Publishing president Sean Moore, publishing director Karen Prince, art director Edwin Kuo, designers Rachel Maloney, Mariel Morris, Gus Yoo, Greg Lum, La Tricia Watford, Erika Lubowicki, senior editor Lisa Purcell, editors Marcel Brousseau, Suzanne Lander, Gail Greiner, Ward Calhoun, Emily Beekman, and Liz Mechem, picture researcher Ben Dewalt, production manager Sarah Reilly, production director Wayne Ellis, and indexer Jessie Shiers; Chrissy McIntyre of Chrissy McIntyre Research, LLC; Wendy Glassmire of the National Geographic Society; Harriet Mendlowitz of Photo Researchers, Inc.

PICTURE CREDITS

PICTURE CREDITS
The following abbreviations are used: PR Photo Researchers, Inc.; SPL Science Photo Library; JI © 2006 Jupiterimages Corporation; SS ShutterStock; IO Index Open; iSP ©istockphoto.com; BS–Big Stock Photos; ARS/USDA–Agriculture Research Service/U.S. Department of Agriculture; NRCS–Photo courtesy of Natural Resources Conservation Service; NOAA–National Oceanic and Atmospheric Association; OAR–Oceanic and Atmospheric Research; NURP–National Undersea Research Program; USFWS–U.S. Fish and Wildlife Service; USGS–United States Geologic Survey; NSF–National Science Foundation; NASA–National Aeronautics and Space Administration; NLM–Courtesy of the National Library of Medicine; SI/NZP–Smithsonian Institution/National Zoological Park; SIL–Smithsonian Institute Library; DL–Dibner Library Portrait Collection; SPS–Smithsonian Photographic Services; AP–Associated Press; LoC–Library of Congress; NGIC–National Geographic Image Collection
(t=top; b=bottom; l=left; r=right; c=center)

Introduction: Welcome to Ecology
IV iSP/Brendan O'Boyle VI JI 1t JI 1b JI 2bl Robert Brock/SPL 2br JI 3l ARS/USDA 3r iSP/Dan Schmitt

Chapter 1: Life on a Small Planet
4 IO/photos.com Select 5t iSP/Daniel Tang 5b iSP/Irene Teesalv 6tl LoC 6b Larry Jacobsen/SS 7tl LoC 7tr Pier deLune/SS 7b LoC 8tl JI 8r Victor Pryymachuck 9t iSP/Martin Pernter 9br Michal Napartowicz 10tl Simon Fraser/PR 10bl Jacques Jangoux/PR 10br JI 11 Mark Bond/SS 12tl JI/JI/SS 12bl James A. Kost/SS 12tr Illustration by Rachel Maloney/Data Source: bbc.co.uk 13 Danis Derias/SS 14tl JI 14b Mark Smith/PR 15tl Nature's Images/PR 15bl iSP/Robert Deal 15tr Science Source/SPL

Chapter 2: The Ups and Downs of Populations
16 IO/photos.com Select 17t IO/Able Stock 17b JI 18tl iSP/Richard McDowell 18tc JI 18tr JI 19l Illustration by Rachel Maloney/Data Source: www.dieoff.com 19tr Illustration by Rachel Maloney/Data Source: www.dieoff.org 19br JI 20tl USFWS/Luther Goldman 20tr JI 21t Gregory G. Dimijan/PR 21bl Stephan Dalton/PR 21br E.R. Degginger/PR 22tl iSP/Mary Marin 22bl JI 22br JI 23 JI 24tl Hans Reinhard/PR 24bl Biophoto Associates/PR 24tr JI 25tl Michael Patrick O'Neill/PR 25tr JI 26tl Eric Grave/SPL 26bl JI 26br Gregory Ochocki/PR 27 iSP/Adam Junrrosi

Chapter 3: Partners in Evolution
28 iSP/Melody Kerchoff 29t JI 29b Muriel Lasure/SS 30tl JI 30b JI 30r Paul Whitted/SS 31tr IO/Amy and Chuck Wiley/Wales 31br JI 32tl BS/Riger 32b Lynn Amaral/SS 33tl JI 33tr Darwin Dale/SPL 33tr iSP/Chart Chai 34tl Krista Mackey/SS 34bl Brad Thompson/SS 34br Alan and Linda Detrick/PR 35tr Dusan Dobes/SS 35b Caitlin Copeland/SS 36tl Uwe Ohse/SS 36b Robert J. Beyers 35 37tl iSP/Mike Tam 37tr iSP/Melissa Carroll 38tl Matt Regan/SS 38r Illustration by Greg Lum/Data Source: www.epa.gov 39t JI 39br © 2006 Getty Images

Chapter 4: A Sense of Place
40 Craig K. Lorenz/PR 41t iSP/Eric Forehand 41b iSP/Jessie Carbonaro 42tl © 2006 Getty Images 42tr Gene Whitaker/USFWS 42b USFWS 43 Tyler Olson/SS 44tl JI 44br Courtesy of Nancy Sefton 45tl iSP/Rene VanderWeerd 45tr Mary Hollinger/NOAA 45b LoC 46tl JI 46bl JI 46br JI 47tl JI 47r John Kaprielian/PR 48tl © 2006 Corbis 48tr Sheila Terry/SPL 48bl © 2006 BrandX Pictures 49t Illustration by Rachel Maloney/Data Source: USGS 49b Digital Vision 50tl Kenneth Eward/PR 50tr JI 50bl Charles D. Winters/PR 51tl JI 51b © 2006 Getty Images 52tl JI 51tr Tierbild Okapia/PR 51br iSP/Cezary 53b Jerry L. Ferrara/PR

Chapter 5: Habitats and Landscapes
54 Brian Hendricks/SS 55t JI 55b USFWS 56tl Digital Vision 56r Nishal Shah/SS 57tl BS/Photobag 57r Steve Maehl/SS 58tl JI 58b iSP/Dustin Belton 59l iSP/Nicholas Belton 59r USGS 60tl © 2006 Getty Images 60b JI 61bl Ra'id Khalil/SS 61tr PR/SPL 62tl JI 62bl Chad Palmer/SS 63tl JI 63r JI 63b JI 64tl © 2006 Getty Images 64tr Bob Williams/NOAA 64bl Gregory Ochocki/PR 65t Dennis Sabo/SS 65b Mona Lisa Productions/PR 66tl iSP/John Marchena 66br JI 67l © 2006 Getty Images 67r © 2006 Getty Images

Chapter 6: Strategy and Self-Sacrifice
68 JI 69t Tischenko Irina/SS 69b JI 70tl Frank B. Yuwono/SS 70r JI 71tl Michael Ledray/SS 71tr Larry Miller/SPL 71cr Brad Phillips/SS 71br Joseph Becker/SS 72tl JI 72b Ruudolf Georg/SS 73t Vladimir Pomortzeff/SS 73b Manoj Alappat/SS 74tl JI 74c J. Norman Reid/SS 75tr JI 75br Cory Smith/SS 76tl EnigmaGraphics/SS 76b Big Zen Dragon/SS 77t JI 77b JI 78tl Tony Campbell/SS 78bl Sinclair Stammers/PR 79 JI 80tl JI 80tr Jill Lang/SS 80br JI 81bl JI 81tr NOAA 82tl IO/photolibrary.com pty. ltd 82c BS/ChickenHound 82b JI 83l Tom McHugh/PR 83r JI

Ready Reference
84t (l-r) Public Domain; USFWS/WV-Carson-Historics; LoC; NOAA/OAR/NURP 84c (l-r) Public Domain; Amy Sussman/AP/JGRYL GRAYLOCK; Yale Office of Public Affairs; SIL/DL 84b (l-r) Mary Evans/PR; AP; International Council for Science; LoC 85t (l-r) Photo by Rick O'Quinn, University of Georgia Communications; Public Domain; SPL/PR; *Croker*, plate following p. 94 85c (l-r) SPL; University of Washington; LoC; SPL/PR 85b (l-r) SPL; Michael Dwyer/AP 86bl (l-r) Ferenc 86br John Arnold/SS 86-87 Illustration by LaTricia Watford/Data Source: © 2006 Conservation International 87tl BS/kiankhoon 87tr Hugo Maes/SS 87cr Chin Kit Sen/SS 87bl Vladimir Pomortzeff/SS 87br Eldad Yitzhak/SS 88t Illustration by Rachel Maloney/Data Source: Woods Hole Oceanographic Institution 88b Illustration by Rachel Maloney/Data Source: Arctic Climate Impact Assessment 89c USGS 89b NASA 90t Illustration by Rachel Maloney/Data Source: NASA/J. Wallace, University of Washington 90b NASA 91t NASA/Jeff Schmaltz/Robert Simmon 91b NOAA/Captain Budd Christman 92 NRCS 93 Illustration by Rachel Maloney/Data Source: Long Island Sound Study 94 Illustration by Rachel Maloney/Data Source: bbc.co.uk 95 Illustration by Rachel Maloney/Data Source: River Bend Nature Center 96tl NOAA/Harold Hudson 96tr ARS/USDA/Jack Dykinga 96bl ARS/USDA/Scott Bauer 97tl ARS/USDA/Keith Weller 97tr ARS/USDA/Scott Bauer 97b NSF 98tl ARS/USDA/Keith Weller 98bl USDA/NRCS/Lynn Betts 99tl USDA/NRCS/Tim McCabe 99cr JI 99br LoC 100tl JI 100r IO/LLC, FogStock 100bl JI 102tl IO/Bruce Ando 102tr FEMA/Adam Dubrowa 102cr Themba Hadebe/AP 102br Szabi Borbely/SS

Chapter 7: Ecology in Geologic Time
102 IO 103t IO/Chris Rogers 103b JI 104tl JI 104tr JI 104bl JI 105tr JI 105br Peter Doomen/SS 106tl Michael Schofield/SS 106bl Eye of Science/SPL 107tl Alan Sirulnikoff/PR 107tr NSF/Dr. Yang Wang/Florida State University 107br Dr. Yang Wang/Florida State University 108tl JI 108bl AP/Matthew Cavanaugh 109t Illustration by Rachel Maloney/Data Source: J. John Sepkoski, Jr., 1997 109b D. Van Ravenswaay/SPL 110tl JI 110b JI 111tl Alexei Dobrovolski/SS 111tr Michael Thompson/SS 112tl JI 112bl JI 113tl JI 113b JI 114tl Romeo Koitmäe/SS 114bl Doxa/SS 114br Alon Othnay/SS 115t SF Photography/SS 115b Benjamin A. Hunter/SS 116tl JI 116b JI 117t Rui Manuel Teles Gomes/SS 117b JI

Chapter 8: What Do Ecologists Do?
118 AP 119t Jostein Hauge/SS 119b JI 120tl Tom Oliveira/SS 120bl JI 120br Jim Jurica/SS 121tl JI 121tr JI 121cr Illustration Research Reserve Collection/NCDDC/P.R. Hoar 122tl JI 122bl JI 122bc JI 122br Barbara Brands/SS 122tr Howard Sandler/SS 123 JI 124tl JI 124bl Lou Oates/SS 124br Simon Pedersen/SS 125c JI 125b Lynn Amaral/SS 126tl JI 126tr Joel Bauchat Grant/SS 126bl JI 127tl JI 127b Shawn Hine/SS 128tl Jim Parkin/SS 128tr Jim Parkin/SS 128b NOAA Restoration Center 129t USFWS/John and Karen Hollingsworth 129b USDA/Tim McCabe 130tl USDA/Keith Weller 130b USDA/Michael T. Smith 131tl JI 131b USDA/Peggy Greb 132tl JI 132b Natalia Bratslavsky/SS 133tl IO/Keith Levit Photography 133tr JI 133bl JI 133br AP/Katsumi Kasahara, Pool

Chapter 9: Survival at Earth's Extremes
134 JI 135tl JI 135b NOAA/Commander John Bortniak 136tl Carnegie Mellon University 136b iSP/Jose Carlos Pires 137l © 2003 National Geographic Society 137r William J. Mahnken/SS 138tl James Behrens (IGPP, Scripps) 138tr Silense/SS 138bl NASA/Goddard Space Flight Center/Scientific Visualization Studio/Canadian Space Agency, RADARSAT International Inc. 139t Silense/SS 139b Silense/SS 140tl Michael Stedinger/Columbia University 140br Illustration by Rachel Maloney/Data Source: Lamont-Doherty

Earth Observatory of Columbia University 141tr Michael Stedinger/Columbia University 141b NOAA/Commander John 142tl NOAA 142bl NOAA/P. Rona 142br NOAA 143 Illustration by Rachel Maloney/Data Source: NOAA 144tl IO/Lauree Feldman 144cl JI 144cr JI 144bl Carnegie Mellon University 144br JI 145 Harry H. Marsh/SS 146 PSHAW-PHOTO/SS 146b NOAA/Michael Theberge 147tl Nathalie Speliers Ufermann/SS 147tr NOAA/Shannon Rankin 147b NOAA/Shannon Rankin

Chapter 10: The Variety of Life
148 Leon Forado/SS 149t JI 149b Hftan/SS 150tl JI 150bl Sasha Davis/SS 150br EcoPrint/SS 151 Illustration by Rachel Maloney/Data Source: © 2006 Conservation International 152tl JI 152b Wade H. Massie/SS 153tr © 2004Coop99/ Darwin s Nightmare 153bl USDA/Scott Bauer 153br USDA 154tl JI 154bl JI 154br USDA/Scott Bauer 155t AP/XINHUA/Wang Chengxuan 155b USDA/Forest Service Southern Research Station/Bill Lea 156tl Alix/Phanie/PR 156tr TH Foto-Werbung/PR 156bl Kathy Merrifield/PR 157 Christine Gonsalves/SS 158tl Cezary Gwozdz/SS 158b AP/Joe Gill/EXPRESS-TIMES 159l NOAA/Jose Cort 159r Michael J. Bryk/SS 160tl Tammy McAllister/SS 160r Brandon Holmes/SS 161t Leon Forado/SS 161b USDA/Resources Conservation Services/Bob Nichols 162tl NOAA/Channel Islands National Marine Sanctuary 162b NOAA/Shane Anderson 163 Illustration by Rachel Maloney/Data Source: © 1995-2005 Tree of Life Web Project 164tl Mark Bond/SS 164rb © 2003 NGIC 165t JI 165b Robert A. Mansker/SS

Chapter 11: The Legacy of Man
166 LoC/Arthur Rothstein 167t LoC 167b JI 168tl Vladimir Korostyshevskiy/SS 168bl Mary Evans Picture Library/PR 168br A&S Aakjaer/SS 169t Illustration by Rachel Maloney/Data Source: William Ruddiman 169b USDA/R. Kindlund 170tl LoC 170b LoC 171t LoC 171b LoC 172tl Phil Barrett/SS 172r Jose Luis Magana/AP 173t Fernando Llano/AP 173b Bill Haber/AP 174tl Jhaz Photography/SS 174bl NOAA Fisheries, NW region, Nick Iadanza 174b Cliff Deputy/SS 175t AP/Damian Dovarganes 175b IO/LLC, Fogstock 176tl iSP/Lawrence Karn 176r AP/Ric Feld 177t AP/Alden Pellett 177b AP/Greg Wahl-Stephens 178tl Jegors Fjodorovs/SS 178c courtesy USGS EROS Data Center and Landsat7 science team. Photographs courtesy Compton Tucker, NASA GSFC 178b USDA/Bill Lea 179tl Sergei Kovalenko/SS 179b IO/John James Wood 180tl Gregory James Van Raalte/SS 180bl NASA Johnson Space Center–Earth Sciences and Image Analysis 180br Galina Barskaya/SS 181t Anna Chelnokova/SS 181b Alan C. Heison/SS 182t BS/hhakim 182c Data Source: NASA/*Climate Change 2001: The Scientific Basis* 182b Laura Rauch/AP 183l Laura Lohrman Moore/SS 183r JD Pooley/AP 184br Darryl Sleath/SS 184cr Toby Talbot/AP 184b NOAA Restoration Center & Damage Assessment and Restoration Program 185t AP/NIPA 185b Laura Rauch/AP

Chapter 12: Ecology and the Future
186 U.S. Geological Survey/Earth Science Photographic Archive 187t Tang Chee Long Gabriel/SS 187b Illustration by Rachel Maloney/Data Source: Walt Parks, University of Georgia Crop and Soil Science 188tl Syed Sajjad Ali/SS 188c LoC 188bl IO/photos.com Select 188br JI 189 Eugene Hoshiko/AP 190tl JI 190cl NASA Earth Observatory 190cr NASA Earth Observatory 191t Natural Resources Conservation Service 191b Thomas Nord/SS 192tl JI 192b Boris Heger/AP 193tl Kent Gidmer/AP 193tr ARS/USDA 193b John McConnico/AP 194tl Dorian C. Shy/SS 194c Yoshi Ogasawara 194b Chen Wei Seng/SS 195tr Earth Sciences and Image Analysis Laboratory, NASA Johnson Space Center 195bl Vera Bogaerts/SS 196tl NASA Earth Observatory/USGS EROS Data Center 196bl AP 196br USFWS 197t iSP/anzeletti 197b Paige Falk/SS 198tl U.S. Geological Survey 198tr Stanislav Khrapov/SS 198br U.S. Geological Survey 199t NSF 199b Risto Bozovic/AP 200tl NASA Visible Earth 200tr NBII Digital Image Library/John J. Mosesso 200b Ron Edmonds/AP 201t Natural Resources Conservation Service/Bob Nicholas 201b ARS/USDA/Keith Weller

At the Smithsonian
208t SPS/Jeff Tinsley [2000-11244.24a] 208b SPS/Jeff Tinsley [90-6258] 209 SPS/James Di Loreto [2003-8959] 210 NASA 211 Photo by Scott Bauer/ARS